*Classicism and Romanticism*

# Classicism and Romanticism

*with other studies in art history*

*by Frederick Antal*

Icon Editions
Harper & Row, Publishers
New York, Evanston, San Francisco, London

First Icon Edition published 1973.

STANDARD BOOK NUMBER: 06-430018-8

# Contents

# Plates

*When not otherwise stated the medium is oil*

## Plates

## Plates

# Acknowledgments

In acknowledging the assistance I have received from several friends in preparing this edition of my late husband's articles, I should like to thank Professor Sir Anthony Blunt and Mr. Arthur Wheen for reading them in their original form and offering most helpful suggestions. I am further greatly indebted to Mr. Arthur Wheen for subsequent help and encouragement and to him and to Mr. Keith Andrews for courageously tackling the translation into English of the article on *The Problem of Mannerism in the Netherlands*, a task which, owing to the complexity of the subject matter, only such perceptive and conscientious scholars could successfully have achieved.

Thanks are due to the editors of the following periodicals for permission to reprint material included in this volume: *The Burlington Magazine* for *Reflections on Classicism and Romanticism* (April 1935–January 1941), *Remarks on the Method of Art History* (February–March 1949), *Mr. Oldham and his Guests* (May 1949) and *Around Salviati* (April 1951), *The Art Bulletin* for *Observations on Girolamo da Carpi* (June 1948), *Kritische Berichte* for *The Problem of Mannerism in the Netherlands* (Nos. 3–4, 1928).

EVELYN ANTAL
*London, 1966*

# Foreword

When Frederick Antal died in 1954, *The Times* remarked that his 'influence on contemporary art history was of great importance in spite of the fact that comparatively little of his writings had been published'. That 'comparatively little' amounted to a single volume, the monumental *Florentine Painting and its Social Background*, published in 1948 when Antal was already sixty-one, and a number of contributions to periodicals, many of them inaccessible except in museum libraries. There were enough to reveal him as one of the most fascinating and original writers of our time, but sadly little in view of the enormous range of his learning. In fact, as *The Times* obituarist implied, Antal left behind him a large body of writings which he had never found time to shape into publishable form. These included his studies on Fuseli and Hogarth, which were fortunately so far advanced at the time of his death that his widow has been able to edit them as he would have wished. But the material which he had long ago assembled for two further volumes on Florentine Painting and its Social Background, dealing with its evolution up to the establishment and first eighty years of the Grand Duchy, is likely to remain in manuscript for ever, the risk of imposing on it a structure unintended by its author would be too great even for an editor intimate both with the subject and with Antal's highly independent turn of thought.

So, although the number of his published books has trebled since his death, they still tell us little, except in the occasional digression, of his discoveries in certain fields of art-history which he had studied profoundly, and in some cases pioneered: for example, the origins and evolution of mannerism, and the interaction of romanticism and classicism, especially from the time of the French Revolution until the death of Géricault. Nor do they state explicitly the principles underlying his art-historical method: they simply demonstrate the consequences of its application to a variety of particular subjects.

The republication here of his more important essays is therefore of the highest value. In most of them he sets forth his conclusions on certain painters and periods with such a wealth of compressed detail that they might almost be synopses of projected monographs. (The Appendix to his *Observations on Girolamo da Carpi* is particularly valuable since it sketches in the background to the projected third volume of *Florentine Painting and its Social Background*.) And in his *Remarks on the Method of Art History* he traces the emergence of his own school of sociological interpretation with a thoroughness belying his preparatory claim that 'the following are a few casual thoughts, in no sense systematic, on the method of art history . . .'. Antal was incapable of the 'casual thought'. Everything he wrote was marvellously precise. Critics who disliked *Florentine Painting* or *Hogarth* did so because they found his Marxism unpalatable or his detail exhausting. They never disputed the exactness of his facts or his observation. (It would have pleased him that a literary critic, Cyril Connolly, should find his *Fuseli Studies* as enthralling as a thriller: he aimed to be as observant, thorough, and objective as a master-detective.) This precision was, in his case, twofold. He 'read' works of art with the care of the professional connoisseur; and he took infinite pains to check that his deductions regarding the circumstances which produced them were not merely plausible but supportable by fact. I remember an occasion when he visited a private collection in Paris to see a Tocqué portrait which he believed must have particularly impressed Hogarth when he visited the *Salon* of 1743, and could have furnished a prototype for his unusually *mondain* portraits of *Mr and Mrs James* (Worcester Art Museum, U.S.A.). This deduction, which he based purely on stylistic evidence, turned out to be fully justified, but with typical caution he referred to it only in a footnote, together with a perfect characterization of Tocqué's portraiture: 'between the emptiness of Nattier and the new Dutch faithfulness of Aved'.

This instance of Antal's tireless quest for accuracy may belong more properly to connoisseurship than to art-history, but it does, I hope, illustrate the care he took to reinforce intuition with fact, and his insistence, rare in less gifted Marxists, on empirically testing every theory, however convenient, attractive, or plausible it might at first appear.

Apart from their sensitivity and historical insight, the essays gathered here are astonishing for their scope and, in the case of those written in the twenties, their modernity. Although Antal admitted, in the *Problem of Mannerism in the Netherlands*, that his conclusions in this, at that time, almost untapped field, were often 'provisional and incomplete' and that 'a thorough examination . . . would elicit important art-historical surprises', this essay is in fact the first serious evaluation of a period worked over today by innumerable specialists. But if, as he anticipated, their researches correct or contradict certain of his attributions, their writings seldom contain much of interest except to each other. Antal had a larger view of the functions of art-history. He was equally familiar with most of the major phases of European art from the Trecento to Expressionism, and it was this, reinforced by his panoramic view of history, which enabled him to accord to artists who had hitherto been regarded as of purely national or psychological interest, e.g. Hogarth and Fuseli, their proper place in the context of European history. It is possible, too, that the freshness of his approach to English art was due partly to his Hungarian family background and continental training. He came to it comparatively late in life and never ceased to enjoy it with an eye undulled by native familiarity.

No aspect of art was beneath his consideration. He often affirmed that minor works could reveal more of a period than masterpieces. Indeed, at one time he found himself fascinated by the illustrations and stories in some Edwardian boys' school books which he found in a country library. These documents of a culture as remote from him as, say, mediaeval Peru's, not only brought him immense amusement but gave him, as they did George Orwell, an unexpected side-light on the taboos and prejudices of the British middle class. I discovered this long afterwards when, after groping unsuccessfully for an answer to a question from him, he astonished me by urging, 'Cough it up, old man!' Unfortunately Antal's humour shows seldom in his writings, for although it played a large part in his character, he considered wit, like literary elegance, irrelevant to objective history-writing.

Yet for all their intellectual discipline, these essays cannot conceal his love for sheer visual beauty. Only a delight in works of art could have inspired him to study them with such unflagging devotion. Marxism has led less sensitive men to judge works of art solely by their content and social origin, often over-simplifying both to ease them into a dialectical strait-jacket. Antal's interpretation of them was utterly free from prejudice. He recognized the seriousness of courtly painters like Bronzino and Watteau and the bathos of much Soviet art. Moreover, his terminology, though never inapposite, is remarkably flexible. If he refused, rightly, to apply the term 'mannerism' to a whole era, reserving it for particular aspects and artists, he also insisted that: 'there is no reason why the term "Expressionism" should be confined to the trend of art which set in on the continent between 1905 and 1910 (mainly in Germany). . . . I do not think the chronological demarcation is of any real importance. "Expressionism" is only a convenient term, like "baroque" or "realism", used in order to define a broad tendency and in no way impeding historical precision' (*Hogarth*, p. 128).

In art-history, as in politics, Antal was systematic without becoming doctrinaire. He knew that abstract principles too rigidly applied can often obscure the truth. This open-mindedness, unusual in men of strong convictions, adds greatly to the forcefulness of his writings and explains how their influence has never been confined by political or geographical frontiers. They dispense, like Shakespeare's King Henry V, a 'largess universal'.

DAVID CARRITT

# 1

# Reflections on Classicism
# and Romanticism

Few conceptions in the terminology of art-history are as vague and indefinite as those of classicism and romanticism. At the same time the methods adopted by modern art-historians to overcome this difficulty clearly reveal the limitations of the purely formal criteria generally applied by them. Thus, a subdivision of the wide and vague stylistic groups of classicism and romanticism has recently been attempted in which the work of certain more limited groups of artists, or even of single artists, is characterized by certain formally more precise terms taken from earlier periods of art. A German writer has described the various styles of French painting during this period—the late eighteenth and early nineteenth century—as follows: the pure 'classicism' of David is followed by the 'proto-baroque' style of Prud'hon and Gros; Ingres' style is 'romantic late classicism, i.e., classicism with gothic and manneristic tendencies'; that of Géricault 'early baroque with realistic tendencies'; and that of Delacroix 'romantic high baroque'.[1] Another art-historian has analysed and classified the German painters of the same period: thus Mengs is called an 'early classicist' by this authority; Fuseli (whose work in this country makes him of special interest to English readers) an 'early gothicist'; the Füger-Carstens group 'high classicists'; another group (Runge, Friedrich) 'high gothicists'; and the Nazarene School (Overbeck, Cornelius, etc.) 'late gothicists'.[2]

From the formal point of view, all these connotations are certainly correct. Nor can there be any objection in principle to the use of terms applicable to the style of earlier, to describe that of later, periods. Certain terms exist in style

[1] W. Friedländer, *Hauptströmungen der französischen Malerei von David bis Cézanne. I.—Von David bis Delacroix*, Leipzig, 1930.
[2] F. Landsberger, *Die Kunst der Goethezeit*, Leipzig, 1931.

*1*

analysis, and it is reasonable to use them for purposes of formal classification. These various definitions, however, do not take us far towards an understanding of the period analysed. Is it really sufficient for the purpose of historical explanation simply to devise a nomenclature for changes of style, to describe the formal criteria of these changes, or to discover formal similarities between the styles of different periods? To state that not only Rubens, but also Delacroix, is baroque is not to explain romanticism. The questions to be answered are surely these: Why are the styles of the eighteenth and nineteenth centuries similar to those of the sixteenth and seventeenth centuries? Why are the artists working about the year 1800 gothic-manneristic, classicistic, protobaroque, high baroque? What is the reality behind these similarities, behind these formal criteria of style? What is the meaning of the conception of style in its totality?

In defining a style I believe that contemporary art-historians frequently devote too much attention to the formal elements of art at the expense of its content. Only too often they overlook the fact that both form *and* content make up a style. If we take sufficient account of content we realize at once that styles do not exist in a vacuum, as appears to be assumed in purely formal art-history. Moreover, it is the content of art which clearly shows its connection with the outlook of the different social groups for whom it was created, and this outlook in its turn is not something abstract; ultimately it is determined by very concrete social and political factors. If, therefore, we wish to understand a style in its totality, we must trace its connection with the society in which it has its roots.

Once we have realized the character of this relationship, we shall no longer exaggerate the significance of formal similarities between the styles of different periods to the extent of using them for purposes of definition. We shall recognize the existence of these similarities, search out their true causes and thus reduce them to their proper significance. Since the same social and political conditions, the same social structure, never recur in different historical periods; since any similarity can only apply to certain aspects of the conditions found; any resemblance between the art of such periods can only be partial. It is true that such similarities can be of considerable importance, if, as is often the case, the structure of the societies in question and their general outlook are fundamentally related to one another. But equally often the

2

'similarity' in style has a very different significance: under greatly changed social and political conditions it may express the outlook of a social group different from that of the earlier period. It is one of the most interesting problems in art-history to explore in this way the basic causes of stylistic similarities and apparent relationships, to show why a certain older style exerts an important influence on a later style in the development of art. The explanation follows quite naturally, almost automatically, when we clearly grasp the changes in the social structure, in the appropriate ideologies and in the styles corresponding to them.

What is the reality behind the terms 'classicism' and 'romanticism' at the end of the eighteenth and the beginning of the nineteenth century?

Let us begin with France and trace the development of French art from its own roots. Where foreign influences—primarily English ones—are important, I shall simply mention the fact. The true causes of such influences— why at a certain historical moment a certain foreign style could exert an influence—will be fully comprehensible only after a study of the development of these other countries and a discussion of the significance of the style in question in terms of its own basic causes. Even in this sketch of the French development, however, I can do no more within the framework of these articles than present the results of my investigation; there is no space to support my conclusions in detail.

I shall discuss first the picture which is generally regarded as the classicistic one *par excellence* of French eighteenth-century art: David's *Oath of the Horatii* (1784) (Plate 21). What significance attaches to the appearance at that moment of a pictorial representation of a heroic scene closely related in its general theme to Corneille's tragedy? What is the significance of a picture so simple and compact in its composition, a picture in which the principal figures are placed soberly side by side, dressed in ancient Roman costume, a picture painted with such severe objectivity? What explains its wildly enthusiastic reception at a time when Fragonard was still painting his sumptuous and whirling rococo pictures, based on the Rubens-Watteau tradition?

It is not necessary to give a detailed account of the well-known political and ideological situation in France at this period. The rising middle class proclaimed new political ideals: democracy and patriotism. It had a new conception of morality: civic virtue and heroism. The pattern of these ideals, as elaborated most effectively by Montesquieu and Voltaire, Rousseau and Diderot, was provided by ancient history. This complex of middle-class ideals immediately enables us to comprehend David's picture. Yet it was painted for the Ministry of Fine Arts? Again, there is no mystery. The Court of Louis XV, even more so that of Louis XVI, was anxious to make some concessions to the new spirit of the middle class and to assume the appearance of enlightened absolutism. It was a reflection of this compromising, yet inconsistent and therefore fundamentally hopeless policy, that the Royal administration should yearly have commissioned, through the Ministry of Fine Arts, pictures with subjects drawn from ancient history, preferably having a moral tendency. The whole tendency of David's painting, however, its exaltation of patriotism and civic virtue in all its austerity, the puritanical economy of its composition, was radically directed against his own patrons. Its overwhelming success, indeed its very existence, were determined by the strong feeling of opposition then prevailing against the demoralized Court and its corrupt government.

This picture is the most characteristic and striking expression of the outlook of the bourgeoisie on the eve of the Revolution. It is rigid, simple, sober, objective, in a word, puritanically rational. Simple groups and straight lines form the whole composition and serve to make it clear and striking. It is the method of composition generally known as classicistic. At the same time there is a great deal of objective naturalism in this painting, a naturalism that determines its sober colouring, its accuracy of detail, its clear presentation of simple objects. This accuracy of detail was the result of a careful preliminary study of the model. A drawing for one of the female figures (Plate 20a) reveals this painstaking exactitude which is almost sculptural. This latter tendency came into prominence at a later, less naturalistic, stage of classicism, a stage in which this style was no longer the expression of the most advanced section of the middle class. In David's early period, however, the sculptural character of the figures was merely the result of his study of the model, a study which distinguished his figures from the picturesque, but often schematic and void, rococo figures. This naturalism of David is as characteristic for the taste of the rising middle class as his classicism; both are inseparable aspects of its objective rationalism. The combination of these two factors in David's *Oath of the Horatii* was the cause of its outstanding success. It must be remembered that naturalism had to assume a classicistic form in a historical composition of this kind in order to be accepted at all by the public of that time. For historical subjects could then only be rendered either in a baroque or a classicistic style. Truly objective naturalism, however, is much more compatible with the latter than with the former. On the other hand, only a historical picture could exercise any far-reaching influence on the public of that period. For there still existed a rigidly hierarchic scale in the social estimation of the several types of painting, a scale dating from the time when the artists had to fight for their social position, for their emergence above the artisan level. In this scale historical compositions were at that time still regarded as the highest category of painting. Classicism, based on naturalism, was thus the historically inevitable style of David's picture, a style which accurately reflected its social background.

Let us for a moment disregard this social background and examine the antecedents of this picture solely as regards its form and its content. Corneille's tragedy is its literary source; Poussin's work its formal one. But in thus evoking the names of Poussin and Corneille we are reminded that David is linked with the art of the period of Richelieu and the young Louis XIV, a period in which the monarchy relied on the support of the new middle class in its efforts to accomplish its centralization programme against the opposition of the nobility. Such a period, characterized by the expansion of the middle class and its rise to a position of power, necessarily produced a classicistic art, i.e. an art arising out of the rationalistic conception of life peculiar to that class.[1] It was the most progressive art possible in the France of that time, just as absolute monarchy with its centralizing rationalism based on middle-class support was the most progressive political régime at that period.[2]

[1] The conception of life, however, of the patrons of the artists is not necessarily the same as that which is expressed in the pictures ordered by them. But it is very significant that the chief patrons of Poussin were high officials, bankers and merchants, and not aristocrats. The case of Le Sueur, the imitator of Poussin's style, is quite similar.

[2] The situation was entirely different in the latter part of the reign of Louis XIV, the period in which considerations of religion determined an irrational policy leading to a grave economic weakening of the country, the period of Madame de Maintenon, the death of Colbert, the revocation of the Edict of Nantes and the persecution of the Protestants, that economically most significant section of the French middle class. The style of painting at this latter period was no longer classicistic, but baroque: Delafosse, Antoine Coypel.

# Reflections on Classicism and Romanticism

Of course the classicism of Poussin and of David are not exactly alike; they cannot be; in spite of their close formal resemblance, in spite of the partial similarity of their social backgrounds. The whole atmosphere of the Court, the outlook on life of the upper middle class, could not, despite its rationalism, be as advanced in the mid-seventeenth century as that of the middle class 150 years later on the threshold of the Revolution. Similarly, in the sphere of art, the classicistic compositions of Poussin differ greatly from those of David in the degree of their naturalism, which is much more the case in David's work. It is this increased stress on naturalism—so characteristic of David and the taste of the rising middle class—which accounts for the fact that the various parts of the composition of the 'Horatii' are not welded into anything like that complete unity which is so characteristic a feature of Poussin's work.

In this way we could retrace the line of development to even earlier stages: we could go back to the classicism of Raphael, the 'inspirer' of Poussin, or to that of Masaccio, the 'inspirer' of Raphael.[1] All these 'classicistic', i.e., rationalistic and highly naturalistic, styles arose at socially and politically progressive moments in the historical development of the bourgeoisie. Raphael's art reflects the secular and national ambitions of the Papacy, the only power with national potentialities in the Italy of that period, and at the same time its most advanced and highly organized financial and bureaucratic institution. The art of Masaccio appeared at the moment in which the Florentine upper middle class reached the zenith of its power.[2]

Thus, in all periods since the Middle Ages in which an advanced outlook arose from advanced economic, social and political conditions of the middle class, a classicistic art expressed the rationalism of that class. But in each period, in each country, we are concerned with a different stage in the development of the bourgeoisie, and the different classicistic styles themselves reflect these different phases. Here I must anticipate my account of the further development by pointing out that this phenomenon—classicism based on naturalism as the style of the upper middle class—is true only of the earlier

[1] Let us take, for example, this line of development: Poussin, *Christ delivering the Keys to Peter* (Bridgewater House)—Raphael, *Christ delivering the Keys to Peter* (Victoria and Albert Museum)—Masaccio, *The Tribute to Cæsar* (Florence, Carmine).
[2] Almost immediately afterwards the power of this class declined rapidly.

phases of the development of the upper middle class; as this social stratum became increasingly powerful after the French Revolution, the naturalistic factor gradually came to suppress, and finally entirely destroyed, the classicistic scheme of composition.

We posed earlier the question, how a picture so surprising as the *Horatii* could suddenly appear in the same period as the sumptuous rococo paintings of Fragonard. The work of Fragonard and similar artists quite evidently reflects the real taste of the Court and of all social circles allied to the Court (especially the wealthiest men in France, the royalist *fermiers-généraux*, one of whom was Fragonard's chief patron). However, David's painting, though contemporary with baroque and rococo pictures, did not arise suddenly as is generally assumed, and as comparison simply with Fragonard would lead one to suppose. Throughout the second half of the eighteenth century, i.e., the time during which the new and revolutionary ideas of the bourgeoisie, the ideas of the Encyclopædists, were ever more consistently developed and widely circulated, the artists were turning increasingly to classicism and naturalism. Suffice it to mention Greuze. His paintings, illustrating moral and sentimental themes of bourgeois life, were important antecedents of David's work. Sentimentality was a middle-class virtue. (It is no mere chance that the eighteenth-century cult of sentiment, just as that of democracy, came to France from the more advanced bourgeoisie of England.) The sentimentality of these pictures was a preliminary stage to the heroism in David's work; it was regarded as fundamentally sincere by the middle-class public who acclaimed it as a reply to the frivolousness and frequent superficiality of rococo painting. Greuze was related to David, as Diderot, who greatly admired Greuze's work, was to Robespierre. The heroism and the much more pronounced naturalism of David was only possible at a time when the Revolution was imminent. But Greuze's work of the sixties and seventies was already pointing in this direction. It is a remarkable fact that the principles according to which Greuze composed his large family paintings were also derived from Poussin. Simple groups and lines, the grave gestures of classicism, determined the composition of these works. But they did not as yet show the concentration later achieved by David. Their naturalism, though intentional, was more superficial. Under the mask of morality much rococo sensuality still appeared in the details. Greuze is typical for the transition from the open immorality of the rococo to the radical, consistently severe, morality of Revolutionary bourgeois classicism.

The names of David, Greuze and Fragonard will now convey concrete impressions of a series of different styles prevalent before the outbreak of the Revolution, styles which reflect the outlook of different classes, or of different phases in the development of single classes. They are an indication of the complex structure of the society of those years, that is to say, of the public for whom the historical pictures were annually exhibited in the Salon. To provide a more complete sketch of the historical and artistic situation of that time, I would mention a further group of historical pictures, commissioned, like the *Horatii*, by the Government. The style of these pictures is of great historical interest, for though basically baroque—the usual style of seventeenth- and eighteenth-century court aristocracies—this style nevertheless approaches the severe principles of classicistic composition (though, of course, by no means as consistently as that of David, the most progressive painter of those years). Moreover, these baroque compositions show naturalistic traits far more markedly than do the usual elegant rococo paintings with their mythological and allegorical themes. This is due to the fact that the subjects are more concrete. Some of these Government orders—and it is these with which we are alone concerned—demanded subjects from the recent history of France. Arising out of the programme of enlightened absolutism, these pictures were required to show the deeds of popular rulers, especially Henry IV. A subject executed by Vincent in 1779 characterizes the mentality of the populace in the days of the Fronde, when angered by the repressive methods of Mazarin and the Queen. It represents an incident on the Day of the Barricades in 1648: Mathieu Molé, first President of the *Parlement*, is urged by the crowd to firmer action to secure the release of his imprisoned colleagues (Plate 22). The style of this picture is undoubtedly baroque. But its essentially modern feature—a feature which even relates it to a certain extent to the tendencies of the David group—is the naturalism of its figures and of its architectonic background. As a consequence, this picture, illustrating an episode from the history of the seventeenth century, painted in a baroque style transfused, and partly suppressed by, naturalistic details, strikingly resembles romantic painting. It looks as if it had been painted half a century later.[1] The same is true, to cite one other example, of a picture for which Ménageot received the commission in 1781. It glorifies another popular monarch since it shows the *Death of Leonardo da Vinci* in the arms of Francis I (Plate 23a). The picture itself is at Amboise, a sketch for it in the Collection Mairet, Paris. The latter shows striking resemblances with romantic painting (especially of the Delacroix imitators), not merely in its composition, but also in its colouring. The Renaissance subject introduces a new colour scheme differing from that of rococo paintings, both by reason of its greater naturalism, and of its use of certain theatrical effects, e.g., the dramatic figure of the doctor in pitch-black in front of the dark-green background.

At this stage it suffices to note the existence, during the Louis XVI period, of pictures showing such striking similarities to romantic painting and to indicate their ideological source. We shall appreciate the reason for this resemblance only after our discussion of romanticism itself.

David also commenced his work in the baroque-rococo tradition. While still a young man, he had been recommended by Fragonard for the task of decorating the house of the notorious dancer Guimard. But during the seventies and especially the eighties, he gradually changed his style. The taste of a large section of the bourgeoisie was revolutionized as social and political conditions approached the crisis; the public now demanded more radical tendencies in art. It is impossible to understand David's development without taking this social background into consideration. Throughout these years he consistently developed towards a naturalistic classicism. He approached this aim by the study of styles of many different periods having the same tendency in common, or at least containing elements that could be interpreted in this sense: ancient Roman sculpture and cameos, early seventeenth-century painting of the Bolognese and Roman schools (Carracci, Reni, Domenichino), the revolutionary art of Caravaggio, recent English engravings with classical subjects, etc.[2] Thus David created the style of the French bourgeoisie on the eve of the Revolution; the *Oath of the Horatii* is the culminating point of this development.

[1] Indeed, when, in 1843, a drawing by Vincent for this picture was exhibited, it had a great success in romantic circles.
[2] Thus an engraving by Conway (1763) after Gavin Hamilton's *Hector and Andromache* influenced David's painting in 1783 of the same subject.

David's political views induced him actively to participate in the Revolution: he was made its art-dictator and belonged to the intimate circle of Robespierre's political friends. The great experiences of the bourgeois Revolution, David's intimate contact with daily events, exercised a profound influence on his art.

We can only realize the great step from the *Oath of the Horatii* to the *Oath of the Tennis Court* (1791, Versailles), if we take into account the official meaning of 'historical composition' at that time. It was a great breach of the convention for an historical painter to take a contemporary occurrence and present it without allegorical trappings as an historical picture (i.e., as belonging to the most highly esteemed category of painting). This David did with his *Oath of the Tennis Court*. We can see even more clearly in this picture that, at this stage of the development, classicism and naturalism were inseparable. A contemporary event of great importance (the oath of the Deputies of the Third Estate that they would not leave their posts) inspired this classicistic-historical painting, and precisely for this reason the naturalistic factor necessarily predominated to a hitherto unheard-of extent. This is so, even though the composition is fundamentally an extension of the classicistic scheme of the *Horatii*. The gestures of the Horatii are multiplied in the gestures of the Deputies as they take the oath, but they are greatly enlivened and have all the tension of actuality. So long as classicism remained in contact with reality, during the French bourgeois Revolution, it necessarily led to a more and more consistent naturalism.

This conclusion is supported not merely by the *Oath of the Tennis Court* or the *Death of Marat* (1793, Brussels)—the deeply moving pathos of which is largely due to the fact that David saw Marat on the day before his assassination, then also in his bath writing down his political ideas, and was able thus to reproduce from memory the position of the dead revolutionary with striking truth—it is confirmed also by David's portraits. While his earlier portraits were still to a greater or lesser degree influenced by the baroque tradition, those of the Revolution period are of a truly stupendous immediacy; they embody the consistent naturalism towards which his historical pictures were leading him. Characteristic portraits of this most fruitful period in David's career, in which art and life were so intimately related are, e.g., the portrait of the well-known Deputy *Barère* (Plate 23b), shown in a spon-

taneous attitude while at a meeting of the National Convention; or that of
*Lepelletier de St. Fargeau* (Plate 20b) the famous champion of the Rights of
Man, surprising in the directness of its appeal and depicting the ugliness
of his features with a frankness that would have been impossible before the
Revolution.

After the fall of the revolutionary Jacobins, representing the interests of
the petty-bourgeoisie, a wealthier stratum of the middle class came to power
under the Directoire. The new fashionable society which opened its doors
even to former royalists, soon turned its back on the severe republican ideal
with its far too puritan standards of morality. This social and political change
put an end to David's political career; it also exercised a profound influence
on his art. The large historical painting he produced during the Directoire
period, he conceived in prison, into which he was cast for his allegiance to
Robespierre. It depicts the scene in which the Sabine women reconcile their
Roman husbands and their Sabine relatives (1799, Louvre). This picture no
longer shows the revolutionary political tendency that was so striking in the
*Horatii* of 1784 or in the *Brutus* of 1789 (David was bold enough to paint the
latter subject instead of the *Coriolanus* with which he had been commissioned
by the Royal administration). If there is any political intention at all in the
*Sabines* it is toward conciliation of the several factions, which was the conservative policy of the Directoire. With the changed outlook, we find a
change of style in the picture. David no longer aimed at concentration and
clarity; he now gave a rather overcrowded compilation of beautifully posed
attitudes. In this picture David's classicism has retreated from the position of
advanced and progressive naturalism he had attained in his historical paintings shortly before and during the Revolution; it has become almost sculptural. His former naturalism has by no means disappeared, but it is confined
to the details and is no longer the dominant feature of the figures, now posed
like statues. There is a tendency towards artificiality and, especially, towards
elegance of line. The nude and semi-nude, barred during the strictly decorous
Revolution, now plays an important rôle. David proclaimed his new style
as a close approximation to Greek art; he began to study the Italian primitives
and Greek vase painting. Taken as a whole, these stylistic features express
the ideas of a fashionable society, preoccupied with the pleasures of life. It
is this combination of joyfulness with a somewhat archaistic elegance which

characterizes David's new style. His portraits during these years reveal a similar change. Thus, the portraits of *M. Sériziat* (1794, Louvre) or of *Mme. Récamier* (1800, Louvre) show an elegant simplification of line which, however, is totally different from the powerful simplicity and concentration, the almost crude naturalism, of the portraits painted during the Revolution. The *élan* of the Revolution has vanished.

Under the Napoleonic régime David was again brought into close touch with contemporary events; he became the Emperor's *premier peintre*. This régime was a compound of rationalist absolutism, military dictatorship, bourgeois rule, liberal legislation, together with the resurrection and creation of an aristocratic court tradition; its art was similarly complex; moreover, it changed its character during the successive phases of the régime. The differences and divergent stylistic possibilities of the various types of painting were to some degree accentuated but, as we shall see in discussing the work of the other painters, they also to some extent converged more than hitherto. The latter tendency is the more significant from the point of view of art-development. David's most important period, when his position was unique among the painters of Europe, had by that time come to an end, nevertheless his work during the Napoleonic period is of great historical interest. He was commissioned by Napoleon to paint several large pictures of important ceremonial occasions in the Emperor's career. The fusion of classicism and naturalism in the *Coronation of Napoleon* (1808, Louvre) is even more intimate than in the *Oath of the Tennis Court*, for the subject was a solemn act of State that required to be presented naturalistically and correctly in all its ceremonial detail at the order of the Court. As a result, only remnants of a classicistic composition remain (especially in the large figures at the sides of the picture); the road towards naturalism was cleared even for the category of large ceremonial paintings. But in the historical pictures dealing with antiquity painted during these years (e.g., *Leonidas at Thermopylæ*, 1814, Louvre) David's classicism has lost its former vitality, it has become rigid and exaggeratedly sculptural. Only the details and the overcrowded character of the composition as a whole reveal traces of naturalism. David, trained in the traditions of the eighteenth century, could not discard the belief in the paramount importance of historical paintings dealing with classical subjects; on the other hand, such paintings could no longer have their former vitality, their immediate topical and progressive appeal in the drab reality of bourgeois existence, once that class had consolidated its dominating position. Hence, the classicism in David's historical paintings dealing with ancient history could no longer be a progressive factor; it necessarily became an obstacle to further progress in art. There is thus a division of David's art, according to its subject, into a classicistic type which is retrogressive, and a naturalistic type which is progressive.

This division becomes even more apparent in the work that originated during his Brussels exile (he was exiled by the Restoration government because he had voted for the execution of Louis XVI). Living in a purely bourgeois *milieu*, the former court-painter produced bourgeois portraits that are among the most naturalistic of any portraits dating from those years. But in this last period of his life his progressiveness is strictly confined to this sphere: there could be no greater contrast than that between the naturalism of his portraits and the lifeless, immobile, even academic classicism of his *Venus and Mars* (1824, Brussels). A great gulf divides the period of David's old age from the Revolutionary period, when he was topical in every sense of the word, when his life and his art, his politics and his historical paintings formed an inseparable whole.

## II

In order fully to grasp what was new in the 'early romantic'[1] baroque pictures of Vincent and Ménageot; to grasp, that is to say, what distinguished them—at any rate in degree—from many similar works of the period, we must thoroughly examine their eighteenth-century roots. Hitherto we have been concerned primarily with the immediate cause which led to their appearance (i.e., the fact that they were Government commissions) and with the naturalistic elements of their style. What then constitutes the novelty of these pictures?

The solution of this problem will be simplified and will, at the same time, yield more satisfactory results, if we extend the scope of our enquiry to a

[1] The tendencies in the art of the eighteenth and roughly the first two decades of the nineteenth centuries that are related to romanticism are usually classified as 'early romanticism' or 'precursors of romanticism'. It is our purpose to make concrete this conception in terms of its social and political background.

slightly later point in the historical scale. We shall, therefore, begin by analysing a picture painted by an artist of the following generation which is clearly related in style, particularly to the sketch by Ménageot, and which shows a similarly striking affinity to the work of the nineteenth-century romantics. We refer to the *Pietà* (Plate 24a) painted in 1787 by the pupil of David, Girodet (Anne-Louis Girodet de Rousy-Trioson, 1767-1824). It is the sketch for an altarpiece painted in the following year by Girodet for a Capuchin monastery and destroyed during the Revolution. It has as motif a grandiose, gloomy composition showing Christ's body, partly raised from the ground and supported by the Madonna. The vast, wild background, rocky cavern, occupies far more space on the canvas than do the figures which hardly fill the lower half of the picture. The opening of the grotto reveals against the horizon the vacant Cross near a pale ray of light. The wildly improvised work is modelled in flaring chiaroscuro. Its dusky emptiness is in many respects reminiscent of Poussin, but the combination of the figures with the overwhelming dimensions of the magnificent, desolate grotto is Girodet's invention. Judged from a purely formal point of view—always an inadequate way of defining a picture—its style may best be described as classicistic, penetrated with baroque elements, as is so often the case in the classicistic painting of the seventeenth and eighteenth centuries. It is the gravity of this work that differentiates it so decisively from even the most highly emotional baroque paintings of pre-Revolutionary decades in France, such as those of Fragonard or Deshayes. Although the latter frequently show the influence not only of Rubens, but even of Rembrandt, they are essentially parallels of the festive, sumptuous work of Neapolitan or Venetian eighteenth-century art, as represented by the painters from Giordano to Tiepolo. Girodet's picture, on the other hand, is distinguished by its gloomy, almost uncanny, atmosphere, and by its markedly naturalistic craftsmanship, which is never found to the same degree in the work of the French artists previously mentioned. Thus we are faced with a picture painted during the year immediately preceding the outbreak of the French Revolution, which yet by means of its deep emotion and moving pathos, seeks to arouse all the religious instincts.

Let us try to reconstruct the ideological roots of this painting. Our earlier discussion of Diderot and Greuze has shown that the outlook of the rising

French middle class, and consequently also its literature, art and aesthetic theory immediately prior to and during the Revolution, were by no means exclusively rationalistic in character. Although the rationalistic elements came to predominate, especially as the Revolution approached, a marked emotionalism was also in evidence, at first largely as a complementary factor, later sometimes in opposition to the former. This emotionalism was, however, by no means homogeneous or clearly defined. It could not have been so, for a certain vagueness, a certain lack of definition is of the very essence of emotionalism, and its content in any given case is in the end determined by the object to which it refers. In the early stages, romantic emotionalism was clearly an expression of the new bourgeois subjectivism, of the new claim freely to show one's emotions and even to describe them, because they were felt to be true. Appearing in progressive middle-class circles approximately since the 1740s, this cult of emotion and sentiment assumed an ever more profound, ever more subjective character in the second half of the century, when above all Rousseau, but of course also Diderot, were influential protagonists. Very often this intensified emotionalism engendered an interest in everything that was exceptional, wild, a mark of genius, fantastic or picturesque, an interest underlining both one's own subjectivism and the uniqueness and individual diversity of nature. People sought in nature a reflection of their own moods, they sought pathos, the macabre, and the sublime. This intense accentuation both of middle-class subjectivism and of middle-class realism usually served—as did the rationalism of the Encyclopædists and in art the naturalistic classicism that emerged from it—to support moral, liberal and often also national tendencies.[1] All these predilections were sometimes combined with a passionate interest for the national past, for the Middle Ages, and in art and literature also for the gothic style.[2] Here also English influences strengthened native tendencies: even the term romanti-

---

[1] This is not the place to describe the degree to which Rousseau was in conflict with the Encyclopædists, or the special position of Diderot within the latter group, or the numerous other cross-currents it contained. But this conflict of tendencies, already in evidence before the Revolution, was of the greatest importance and its significance increased in later periods, when only those ideas were selected and adapted from the wealth of eighteenth-century thought that answered the specific purposes of the time in question.
[2] Thus Rousseau linked the emotional with the gothic. He described the letters of the *Nouvelle Héloïse* as being 'in the Gothic manner'.

cism, which is usually applied to all the manifestations we have enumerated, was imported from England.[1] Even religious sentiments were comprised within this whole complex of ideas. Though, as we should expect in progressive bourgeois circles of the second half of the eighteenth century, it was the universal and emotional aspect of religion that was usually emphasized, not its ecclesiastical form: Rousseau provides the best example.

So far, these manifold elements of emotionalism—in common with all other ideological phenomena—have only been examined within the self-contained limits of the history of ideas. As a result they have been reduced to the common denominator of 'early romanticism' or 'early gothicism'. But in the writer's opinion it is more correct to state that during the pre-Revolutionary period a whole series of different emotionally charged conceptions arose within the sphere of the advancing middle class, and that in the course of development these conceptions tended to be blended in an infinite variety of gradations with the existing ideologies of the various social and political strata and consequently also with the manifold movements in literature and art. For the approaching Revolution led to an ever more pronounced differentiation between the several social strata and their ideologies. Thus, for example, in middle-class circles with vaguely progressive or moderately liberal ideas, having no real desire for a revolution, the drabness of daily existence and its actualities gradually tended to lose their emotional appeal and to be increasingly replaced by 'interesting', macabre, fantastic, and frequently also by religious, ideas, religious sometimes even in a more ecclesiastical sense. On the other hand, emotionalism could organically enhance the stern sobriety of the rationalistic and classicistic ideology characteristic of more consistently revolutionary sections of the bourgeoisie. So, on final analysis, we are left with different phenomena—ever more so as the Revolution approaches—even though certain aspects of these phenomena may be strikingly similar.

If we remain in the sphere of art, and more particularly in that of pre-Revolutionary French painting, early romanticism appears characteristically

[1] Letourneur explained this term in the Introduction to his Shakespeare translation (1776) as follows: 'Les tableaux de Salvator Rosa, quelques sites des Alpes, plusieurs jardins de campagne de l'Angleterre ne sont point romanesques, mais on peut dire qu'ils sont plus que pittoresques, c'est à dire touchants et romantiques.'

enough in this sphere only as an infusion, as a subsidiary element within some other style, the character of which is thereby changed.[1] The emotional element could be combined in numerous gradations and in varying strength with the most heterogeneous artistic currents: with, for example, the rococo art of the court aristocracy,[2] the baroque historical painting of the moderately liberal court and middle-class circles, the bourgeois sentimentality of Greuze, or the naturalistic classicism of the most progressive strata.

It is, in fact, merely a matter of varying processes of intensification, emotional modulation, or gradation of already existing tendencies. Emotional appeal, intensified beyond the sentimental towards the interesting and exciting, and, closely related to this, an equally exceptional degree of naturalistic craftsmanship and local colour account for the extraordinary similarity between Ménageot's baroque sketch (Plate 23a) and romantic work of the nineteenth century. A number of similar patriotically royalist paintings of the pre-Revolutionary period also representing scenes from the older history of France in a baroque-rococo manner, for instance, those of Fragonard dealing with the life of Henry IV, or the various pictures of Brenet and Beaufort on the deeds of Bayard (Plate 24b), show nothing like the same degree of affinity. In the classicistic style the new emotional element could appear in a similar manner as a merely complementary factor of varying intensity: it could amplify the stately grandeur of classicism with an element of excitement and fantasy, preferring unusual and interesting situations. Even David did not escape this influence in spite of the fact that his style was

[1] 'Early romanticism' is an international phenomenon. Here we are only concerned with France. The social and political conditions of England or Germany were entirely different. Specific reasons prevented the occurrence of the bourgeois revolution in either of these countries at that period and as a result the tense expectation and excitement of the pre-Revolutionary middle class had to find an outlet in an intense, aggressively individualist 'romantic' movement that came more and more to discard its naturalism and to accentuate the irrational. In those countries 'early romanticism' comprised several different styles of its own according to the successive phases of development. This is particularly true of the successive phases of Fuseli's work which links the artistic development of Germany and England.

[2] The unacademic court style of the eighteenth century from Watteau through Boucher to Fragonard was also to a certain extent emotional, anti-rationalistic. Its theorists continually spoke of 'cœur' and 'sentiment'.

becoming more and more classicistic. Thus he included Fra Angelico and mediaeval sculpture among his studies and painted a *Death of Ugolino* one year after the *Horatii*.[1] Finally, to return to the Girodet picture with which we commenced our discussion, the traditional baroque-classicistic type of religious painting commissioned by the Church could also be subjected to this 'romantic' influence which transfused it with a new atmosphere of emotional tension and pathos. Where this occurred, it signified a greater profundity and strength of religious painting and marked an attempt to find new roots for it. David would certainly have refused the commission for an altarpiece in 1787. His last (and only) work of this kind, a Caravaggesque *Madonna with St. Roch* is dated 1780, and this interval of seven years during the decade preceding the Revolution was of decisive importance. It determined the difference in the outlook of the two artists,[2] and consequently also the diverging paths along which the art of David and of Girodet was bound to proceed. The naturalistic and classicistic development of the progressively minded David was the subject of the preceding section. Let us now trace the development of Girodet, who is usually labelled a classicistic and early romantic painter. It will serve as an instructive counterpart to David's classicism and will show also which early romantic elements were selected during the successive phases of the political development in France by a somewhat conservative, and at the same time talented, artist.[3]

[1] It is interesting to note that conservative devotees of baroque painting attacked even David's naturalistic classicism as 'gothique'; he and other painters following a similar aim were accused—the parallel to Rousseau is obvious—of wishing to take art back to its infancy. The habit of the David circle to appeal to the dry, objective naturalism of the Quattrocento led the author of the Salon review *Les Elèves au Salon ou l'Amphigouri* to complain in 1789; 'N'a-t-on pas eu l'insolence de m'injurier en pleine Académie à cause que j'ai eu la bonhomie de dire que j'aimais le Cortone et ne pouvoir souffrir le Perugin.'
[2] Even in 1789 Girodet made a drawing for a Bayard subject, the usual theme of the patriotic royalist painters of the pre-Revolutionary era. David, on the other hand, painted already in 1781 a picture of Belizar as a blind beggar (Lille), this typical symbol of princely ingratitude which from that time, became very popular in revolutionary middle class circles (moreover, it is very significant that it was with this picture that David took the decisive turn toward his later consistently severe classicistic style).
[3] I give a relatively more detailed account of Girodet than David, because hitherto the former has received little attention from students, and the facts here presented may therefore be of interest, even apart from their re-interpretation.

Girodet's art during the Revolution period, most of which he spent in Italy, diverges completely from the consistent classicism with which David supported the Revolutionary cause. Even if, in 1792, he did paint a moralizing classicistic picture—though characteristically it is at the same time a collection of psychologically interesting character-studies—showing Hippocrates refusing the presents of Artaxerxes (Paris, École de Médecine), he simultaneously produced such a work as his *Endymion* (1791, Louvre), a politically irrelevant, mythologically interesting picture, in which the contours of the elongated figures are bathed in light in a manner calculated to create a mystic, dream-like atmosphere, a thing inconceivable in David's work. Yet it was precisely this that Girodet was most anxious to achieve: to be different from his teacher, and in general to be original and new in his art. He was obsessed with the desire to be original; such an aspiration was fruitful in the original setting of early romanticism, when it expressed the subjectivism of the advancing middle class; but inevitably it tended to degenerate into a glorification of the extravagant, the out-of-the-ordinary for its own sake, when in later, politically less conscious, periods it became purely an end in itself.

After his return to Paris during the Directoire Girodet, himself a very wealthy man, became the favourite of the pleasure-hunting *nouveau riche* society. Although at this stage political power was essentially in the hands of the wealthy bourgeoisie and speculators, the Royalists, who felt themselves safe again, set the fashion and determined the outlook of the social world. There was, therefore, an inevitable reaction against the moralizing, and at the same time severely naturalistic, classicism of the Revolutionary period. We have seen how David reacted to this new situation by emphasizing the artistic, the nude, the beautifully posed, the elegance of line, and the exaggeratedly sculptural, how at this period naturalism ceased to be the dominating note of his classicistic style.[1] With the conservative, anti-Revolutionary and more intellectually snobbish Girodet, this change towards piquant and interesting motifs was naturally far more pronounced. His *Danæ* (Plate 27a) of 1799—contemporary with David's *Sabines*—is a characteristic example of his work during this period. A well-known actress of the

[1] In 1799 Guérin painted a *Return of Marius Sixtus* (Louvre), a more stiff and wooden rendering of David's classicism. But since its theme alluded to the return of the aristocrat emigrés, it met with an immense success.

time, Mlle Lange, had refused a portrait of herself by Girodet, who avenged himself by painting her in the rôle of avaricious Danæ. At her side her lover, a banker, is shown as a turkey. She herself is entirely nude and the erotic intention of this nakedness is emphasized by her sole garment: a ribbon and heron's feather in her hair. Contrivances such as this are, of course, characteristic of the mannerist painting of the sixteenth century. The sensitive lines and elongated proportions, the smoothness and exaggerated punctiliousness of detail point towards the same mannerist influences, e.g., of Parmigianino or Bronzino. The tendency to imitate the calculatedly artificial and essentially anti-naturalistic mannerist style, to imitate, that is to say, the art of the reactionary courts of sixteenth-century Europe, is apparent not only in Fuseli's English work, but also in the French painting of the Directoire period. The mannerist adaptation of classicism in pictures such as Girodet's *Danæ* indicates a tendency to return to the erotic content of eighteenth-century court painting. (This again shows how relatively useless are purely formal criteria for the purpose of defining a style.) The decisive transformation of the social and political constellation of French society since the Jacobin era, the reactionary tendencies of the later phase, are reflected in the anti-naturalistic features of pictures of this type. The same significance attaches to the fact that 'interesting', sensational, piquantly erotic elements were at this stage derived from the early romantics or, perhaps more correctly, it is that romantic elements were transformed and further developed in these directions to such an extent that they appear as the direct opposite of the original Rousseau and Diderot current. If at any one time the early romantic tendencies of differing groups were found to differ in their significance, this was even more true of the romantic movements of the successive régimes.

Another aspect of early romanticism yet more clearly connected with the original movement, or rather with the emotional and sentimental elements of that movement is illustrated by Girodet's picture of 1801 showing Ossian, Fingal and other Celtic heroes receiving Napoleon's fallen generals in their cloud palace (Plate 26). The picture was painted for the Consul Napoleon Bonaparte and was destined for his house in Malmaison. The unparalleled enthusiasm with which Macpherson's *Ossian* was received in the France of the Rousseau period is clear proof of the prevalent interest in everything fantastic, uncanny, mediaeval, or related to the dawn of the national past.

These interests were even more fashionable in the Napoleonic era—*Ossian*, and also *Werther*, were among the Emperor's favourite books.[1] In Girodet's picture the virgins, bards, and heroes emerging from the mist are shown side by side with the Napoleonic officers in their modern uniforms, a combination of naturalism and anti-naturalism highly characteristic of the Napoleonic age, the complexity of which we have already indicated. The semi-nude, rhythmically stylized women show some influence of Flaxman's ultra-classicist, severely simplified linear style.[2] The greater sensuousness of the figures (corresponding to the pre-Flaxman line of English art represented by Fuseli with his mannerist tendencies), which Girodet added, was due to the peculiar outlook of Directoire society and did not disappear under the Empire, except in the work of the consistently naturalistic and progressive artists. The statuesque, excessively precise and, in their details, accentuatedly realistic officers,[3] bathed in 'mystic' rays of light can indeed, hardly be said to conflict with the nebular phantom figures. But the complex Napoleonic age called also for other, more essentially naturalistic modes of expression. How great, for instance, is the difference between these soldiers of Girodet and the consistently naturalistic, consistently pictorial, soldiers painted in the same year by Gros in his sketch for the *Battle of Nazareth* (Nantes). They and Gros' later paintings (*Napoleon in the Plague Hospital of Jaffa*, 1804; *Napoleon on the Battlefield of Eylau*, 1807, both in the Louvre), also of course David's work of this time, led straight to the art of Géricault. Girodet tolerated naturalism

[1] Another picture, painted by Gérard for the same room in Malmaison, has a similar theme. It shows Ossian singing to the accompaniment of his harp in the moonlight at the foot of a ruined gothic castle; as he sings, the characters of his poems appear phantom-like. This is probably the picture now in the Berlin Schloss. (Plate 27b.)

In this connection it is worth recalling that Josephine wished to have a real gothic ruin for her Malmaison garden. In fact a whole portal from a monastery was to have been brought there from Metz.

[2] The influence of Flaxman's engraved book-illustrations is of course even more pronounced in Girodet's illustrative work, especially for the Aeneid; it is also more evident in a sketch to this picture (Louvre) than in the finished work. Like Flaxman, Girodet was a careful student of Greek vase-painting and its technique of colour drawing. The ultra-classicistic style of the early Ingres, which is related both to Flaxman and to Greek vase-painting, will be discussed later.

[3] David censured the figures of this painting as crystallic. There could be no more just description also of mannerist figures.

only in details. It was his aim that the effect of the whole should be exciting, tense, bizarre but above all, as he himself wrote in a letter to Bernardin de St. Pierre, novel. He claimed complete originality for his *Ossian*, while holding that in all his previous work he had still been under the influence of antiquity or of nature.

Girodet developed more and more towards that which he conceived to be original, but which was in reality only what conservative intellectuals of the Empire considered so. Thus, while his *Flood* (1806, Louvre), with its rising waters, its bare rocks, shattered trees, lightning flashes, and its figures which are essentially translations of Michelangelo into a language of wild excitement and horror, already strongly suggests Chateaubriand, that writer was indeed the direct thematic source of Girodet's next large picture, the *Burial of Atala* (1808, Louvre). Chateaubriand's Atala novel, Girodet's Atala painting are clear symptoms of the same tendency; a religious revival.[1] Napoleon and the upper middle class—the Empire had brought the final victory of this stratum —again required the help of the Church. The Concordat was signed and the cult restored (1802).[2] Thus the field was clear for a development of those elements of early romanticism which were tinged with religious emotion and capable of being transformed into religious feeling in the ecclesiastic sense. But however much the protagonists of this new tendency might take from Rousseau, they had necessarily to abandon his point of view as a whole. Chateaubriand, who turned his back definitely on eighteenth-century rationalism, created the ideology for this early romantic ecclesiastically religious revival. The chief claim he made for Catholicism was that the fullness of its beauty was beneficial for art.[3] This was advanced as the new, decisive argument for the intellectuals who, having lived through the age of the Encyclopædists and the Revolution, had ended by becoming politically indifferent and generally disillusioned. To return to Girodet: his *Pietà* of 1787 showed that a combination of early romantic with ecclesiastic elements was to a certain

[1] The fact that Girodet painted Chateaubriand's portrait (Versailles) is further evidence of the close contact between the two.
[2] In actual practice, though not officially, the latter had already been restored under the Directoire, since 1795.
[3] He was enthusiastic about gothic cathedrals, as was Girodet in his didactic poem *Le Peintre*.

degree possible even in pre-Revolutionary France. During the Empire, Girodet was at last able openly to show his earlier loyalty to a Church, which had again obtained official recognition. The romantic revival of religion under Napoleon was, of course, much more reactionary in character than that on the eve of the Revolution. In 1787, it expressed the attitude of the liberal wing of the conservative and ecclesiastical groups[1] who were struggling hopelessly to stave off the revolt by making concessions to the left.[2] But in 1808 the romantic return to religion was a sign of the growing reaction against the achievements of the French Revolution. It could now even become a tool in the hands of those who—like Chateaubriand—were beginning to advocate a restoration of the aristocratic and legitimist régime of the Bourbons. Apart from the religious appeal, however, both Chateaubriand's novel and Girodet's picture contain a further essentially romantic element. Rousseau's enthusiasm for the *pittoresque*, free, unspoilt nature had by this time—Bernardin de St. Pierre provides the literary link—been transformed into an interest in the exotic, strange and macabre. Chateaubriand created an exotic atmosphere of Catholicism, populous with creatures of lyric emotion, tinged with sensualism. The unspoilt 'homme de la nature' of Rousseau has become the Christian Red Indian of Chateaubriand and Girodet. Atala, the half-caste betrothed of the young Indian Chactas, poisons herself rather than yield to her lover and so lose her chastity, since her Christian mother had dedicated her to the service of the Virgin Mary. Girodet depicted the scene in which Chactas and an old hermit put Atala in her grave (Plate 25). The artist treated this motif as a combination of the two themes of The Lamentation and Entombment of Christ. Thus it is not surprising that the *Atala* picture is in essence a mere variation of the *Pietà* he painted twenty-two years earlier; the formal motifs of the figures and their poses, the manner of their composition, and their expressions are strikingly similar in the two works. The same is true of the scenery with its empty cavern, through whose entrance the Cross is again revealed on the horizon. Never-

---

[1] Rousseau's pupil Bernardin de St. Pierre whose *Paul et Virginie* was written in the same year in which Girodet painted the Pietà sketch, is a typical representative of this religious, though not ecclesiastic, left wing. He was equally opposed to the atheists, the scientists, the wealthy, and the aristocrats.

[2] In the opening stages of the Revolution, in the États Généraux of 1789, the clergy were even allied with the middle class against the aristocrats.

theless, the later work shows a marked decline of religious intensity as compared with the very early sketch (nor is this decline entirely due to the difference between the spontaneity of the sketch and the laboured smoothness of the completed picture). Girodet's object in the later picture—just as that of Chateaubriand—was to prove the artistic potentialities of the Catholic faith. Its stage effects are more artificial and its poses theatrical; the sincere spiritual impetus of the *Pietà* is lacking. Lacking also—though not entirely— is the solid pictorial naturalism of the earlier work, or of the paintings by David and Gros depicting contemporary events. And of course the open sensualism of the Directoire Danæ is also absent. The whole—a perfect embodiment of the aesthetic neo-Catholic ideology—is a mixture of religious display and veiled sensuality.

Girodet found further opportunity to show his preference for the exotic when he was commissioned to paint the *Revolt of Cairo* (1810, Versailles), one of the numerous pictures glorifying the deeds of the Emperor. Napoleon's oriental campaigns provided the artists with an opportunity of combining romantically exotic with topical elements.

Things were different under the Restoration for which both Girodet and Chateaubriand had to a certain extent cleared the way in the ideological sphere.[1] There were no more eastern campaigns. If Girodet still wished to paint the exotic, he had to invent a historical theme with an additional royalist and Catholic appeal. Thus he planned to paint St. Louis as a captive in Egypt. He projected a whole series of similar royalist compositions in this, his latest phase. For during the Restoration, when David was exiled, Girodet was one of the most fashionable artists. It is certain that then at last he felt himself in his true element, more so than during the Directoire, more so than during the Empire. The time had arrived when his large compositions could be fully appreciated. They were acquired for the Louvre (among them of course his *Atala*) and Girodet painted full-length portraits of Cathelineau and General Bonchamp, former leaders of the Royalist counter-revolution of the Vendée.

[1] Chateaubriand served as a minister and ambassador under the Restoration, but not uninterruptedly or wholeheartedly. The strong survivals of his early romanticism, his accentuated subjectivism prevented him from identifying himself without reservation with the official love for order and security.

As we have seen, Girodet's outlook was clearly conservative even before the outbreak of the Revolution. The general trend towards reaction—at times open and precipitated, at times less apparent—that marks the interval between the fall of the Jacobins and the Restoration, made him move more and more to the right. This change in his outlook expressed itself in his art as a tendency more or less to abandon the progressive naturalistic basis of the classicistic style. He absorbed, or gradually transformed in this anti-naturalistic sense, numerous elements of early romanticism: religious, erotic, fantastic, exotic elements. We do not, of course, deny the naturalism of Girodet's art. But it is not a consistent, progressive naturalism. The degree and manner in which Girodet ceased to be consistently naturalistic differed in the successive political and ideological phases of his age. During each phase he painted what a sensitive, conservative artist, who at the same time wished to be original and extravagant in his ideas, was able to give to the existing régime. His art excellently illustrates the changes which the progressive bourgeois tendencies of early romanticism could undergo in each successive régime.

### III

We have now examined the many-sided character of the main artistic tendencies in France at the end of the eighteenth and beginning of the nineteenth centuries, vaguely and intermittently designated 'classicist' and 'romantic'. Our attempt at clarification has been furthered by a consideration of the social factors and the political ideas which formed the background of these artistic currents. Using the same method already applied to David and Girodet, we now proceed to analyse the style of the greatest French artist of the early nineteenth century, Géricault (1791-1824), who has usually been called, in a similarly indefinite way, 'romantic'.

The first years of Géricault's artistic activity fall within the last phase of Napoleon's rule, while his main activity is covered by the Restoration, the reign of Louis XVIII. During Napoleon's régime, the complex character of which we have already summarized, the middle class was able to expand economically.[1] Though some liberal and democratic measures were intro-

[1] Hindered only by the element of insecurity, more apparent towards the end of the period, which resulted inevitably from continuous wars.

duced from above, or retained as legacies of the Revolution, the champion-
ship of liberal ideas from below was not encouraged during this absolutist
régime; indeed, the bourgeoisie, being economically and, to a degree, socially,
satisfied, did not lay great stress upon their promulgation. When, however,
under the Bourbons, the aristocratic *emigrés* returned as the new ruling class,
the bourgeois owners of estates confiscated at the time of the Revolution
were in constant dread of losing their property; then, too, their advance,
both socially and politically, and the dissemination of liberal thought through
press and speech was hampered at every turn. Hence it came about that, with
a middle class no longer at ease economically, socially or politically, and
desirous of securing the position due to their strength, liberal opinion took
on a much more pronounced flavour than under Napoleon. A most active
and passionate liberal opposition grew up which, in order to make its struggle
more effective, allied itself with the ill-used, discontented soldiers of the old,
Napoleonic army.

How does the idea-complex, termed romanticism, with its multiple, often
contradictory, elements, signifying something different in every generation
and in each political camp, fit in with these events and ideas of Géricault's
period? We have spoken of the origin of romanticism within bourgeois
mentality as early as the middle and second half of the eighteenth century.
We have seen how the cult of emotion and sentiment, an interest in the extraordinary—ideas called 'romantic' even at that time—were then connected
with middle-class subjectivism and so related (especially in Rousseau and
Diderot) to more or less progressive thought. Further, we noticed how later,
under Napoleon, emotionalism as well as curiosity about the strange and
exotic became fused with rather more conservative ideas—religious and
mediaeval, chivalresque ones—which assumed actuality through the restoration of the cult and the religious revival following it. In Napoleon's day,
this whole world of 'romantic' concepts was to a great extent (Chateaubriand) connected with aspirations for a Bourbon Restoration. After this
had been realized, many conservative intellectuals and others, lacking interest
in, or disillusioned by, politics, clung to these very dynamic 'romantic' ideas,
reacting in part against the cultural deadlock of the Napoleonic period, with
its stagnant classicism, once revolutionary but now devoid of *élan*. Most
writers of the time (Lamartine, de Vigny, Victor Hugo) either attached to

the aristocracy or else living, with their predilection for the extraordinary, in their own ivory tower of a romantic past, long remained partisans of the Bourbons. Stendhal was perhaps the first writer of the opposition to regard 'romanticism' as the literary pendant of liberalism; yet by romanticism he understood no vague portrayal of the passions but a most precise, as it were, scientific delineation of them, not at all incompatible with the simplicity and clarity of the classicist style. When Bourbon rule was found to have many more retrograde tendencies than Napoleon's, serious scholars, attached to the middle class and middle-class culture, very soon turned against it; only by slow degrees, mostly as late as the particularly reactionary reign of Charles X, did any romantic writers, outstandingly Hugo, also join the liberal opposition.

How then does Géricault, usually called one of the most characteristic 'romantic' painters, with his liberal outlook, his singular interests, his passionate, yet consistent, artistic temperament fit into this picture? Under Napoleon, that is, in his early youth, Géricault did not reveal any particular interest in politics. Yet just as the bourgeoisie itself, at the beginning of the Bourbon régime, entertained hopes of greater political freedom, so he too would seem to have felt a certain sympathy with the reigning house. Having always been drawn towards soldiers and the soldier's life, after Napoleon's fall, when only twenty-three, attracted by their striking red uniforms, he could not resist the 'romantic' temptation of joining the fashionable *Mousquetaires Gris* founded under Louis XVIII. At the time of Napoleon's landing even liberal students demonstrated against his return to the throne. But later, when this seemed inevitable and Louis XVIII was deserted even by most of his own soldiers, Géricault, with his impulsive temperament and his quick sympathy for the outraged and betrayed, accompanied the King to facilitate his escape. Nevertheless, this partisanship for the Bourbons, comprehensible even from a liberal standpoint, was only a passing episode in Géricault's life and one he preferred to ignore in future years. Very soon, much sooner than romantic writers of the time, Géricault went into intense opposition to the régime of the Restoration. For, as both his works and his letters show, he now became, once again ahead of these intellectuals, much preoccupied with politics. Although his views were by no means clear-cut, they were nevertheless as decidedly liberal and humanitarian as those of the most progressive section of the middle class of his time. The liberalism of such scholars as Guizot, Cousin, Villemain, Thierry and Thiers, showed very different complexions, and if they were all to a greater or lesser extent against the Bourbon régime, not all of them were as radical as Géricault. His was a militant, not a passive, liberalism, alike in home and foreign policy, which brought him up against the ruling reactionary circle and the clergy, allied to the Bourbons. Like all liberals (Béranger, for instance, who frequented the same social set), Géricault's sympathies were with the Napoleonic soldiers and, in the wider sphere, with nations fighting for their independence, for oppressed and despised races, for all who suffered injustice, even—exceptional for liberals of those years— for the poor. Géricault's outlook led not only to the revolution of 1830, to the overthrow of the Bourbons; despite his early death, he was—as Focillon rightly puts it—already the man of '48. Compared with more genuine 'romantics' such as, for example, Delacroix and even Byron, his political sympathies and antipathies were much more consistent and of a much more radical nature.

Yet Géricault's liberal and humanitarian standpoint was in no contradiction to his stirring, volcanic, somewhat extravagant temperament. Admittedly such a temperament is in itself usually called 'romantic', but we must of course enquire into the objectives and interests towards which Géricault's temperament led, and how he pursued them. We do not find in him the hero-worship so common in romantic types, fundamentally reactionary or with only a slight liberal tinge—his sincere humanitarian outlook would never admit of this. Nor had he, to any extent, the conservative's romantic interest in the Middle Ages.[1] With all his energy, he was immersed in the whole moving life around him. Temperamentally attracted by exciting colourful things, he loved the world of soldiers, the circus, horses, races. To these—so close to life, with nothing of the supernatural about them—his art was to be largely devoted. Yet he did not lose himself in emotion, but like

---

[1] Géricault made only a very small number of drawings with medieval themes, although this period was of general interest to romantics in both political camps. In this respect, Géricault reminds one of Stendhal, his fellow liberal. Stendhal, though making a study, in his own precise way, of men's passions in the Middle Ages, did not hesitate to give expression in this respect to an 'anti-romantic' point of view ('Un nouveau quinzième siècle est impossible').

Stendhal approached his subjects in a scientific, objective frame of mind, studying and analysing them with a concentration and tenacity, previously inconceivable. Nevertheless, it is not enough to assert that Géricault's artistic interests were of a more realistic than 'romantic' character, we must also seek the basis or at least the germs of them in various artistic trends of the Napoleonic era. For, by keeping constantly in mind his link with tradition, with the art of the past, his personal achievements, his thematic innovations as well as his manner of expressing them, stand out the more clearly.

Géricault's choice of masters is characteristic of his whole mentality and of his novel attitude to the various sections of contemporary art. Unlike other ambitious artists of the time, he did not find his way straight to the great historical painter, David, nor even to one of his pupils, but to a painter who represented a genre much lower in the artistic hierarchy. Since he was himself much interested in horses and races, and estimated their representation quite as highly as historical pictures, he did not hesitate to choose as his first master, Carle Vernet, known for his small genre pictures and engraved drawings depicting contemporary society scenes, horse races, dances, even mere fashions. It might have been expected that Géricault, following current practice, would have continued in the line of his teacher, a minor artist not aspiring to the laurels of historical painting, then considered the highest branch of art. Instead, having learned all he could in Vernet's workshop he then became a pupil of Guérin, one of the most celebrated historical painters of the time, himself a faithful follower of David. Thus Géricault, for all his innovating spirit, did not reject tradition nor repudiate the customary David training, but accepted the official, classicist art of the Napoleonic period. In this there was no contradiction. For just as he realized that official and artificial distinctions within art are irrelevant, so he would seem to have felt that classicism and truth to nature are not necessarily in conflict, and that classicism, to remain progressive, as at the time of the French Revolution, must be based on naturalism, on an ever-increasing naturalism. Géricault's first sketches, done in the workshop of Guérin, are those of a typical David follower: the usual antique themes in the accepted manner (even if the figures are full of action). But it is in keeping with all that has been said about him that, even from the period of his studies under Guérin, he was copying, not only the usual masters of the Raphael-Poussin current, but also painters, less familiar at the time, such as Titian, Veronese, Correggio, Caravaggio, Rubens, Velazquez, Rembrandt, Carel Fabritius. Thus from the very first are apparent numerous potentialities dormant within him, far surpassing a conventional, dry classicism. Undistractedly holding on his way, developing organically, he was always to be capable of choosing for his aid such works of art, officially recognized or not, as were necessary for his future development.

Géricault's earliest pictures between 1812 and 1814, in the last years of Napoleon's reign, are of military subjects. In many respects, they certainly bear a close resemblance to the usual works of the kind in which Gros, David's pupil and the most renowned French battle-painter of the period, used to excel. Nevertheless, there are outstanding points of difference. If Gros, for official purposes, placed Napoleon at the centre of the scene or painted the portrait of this or that famous general, Géricault, as a free agent, democratized the representation of the Napoleonic wars by changing the motifs; his 'hero' is a single unknown officer, even a simple *cuirassier*, forced by his wounds to quit the battlefield (Louvre).[1] In the past, only engravings or small-scale paintings ranking as genre pictures could claim such themes as Napoleonic troops fighting without the Emperor, or individual, rank and file soldiers of his army. Géricault not only portrayed such 'democratic' themes with a consistency hitherto unknown, but even painted them on a scale previously reserved only for historical pictures. Indeed, the social gulf between the different types of painting—a legacy from the days when artists had still to strive for a social position above that of artisans—began to lessen with Géricault. The democratization of middle-class society had reached a point at which it was now no longer necessary for artists to pose as historical painters in order to enter its respectable circles. Géricault was the first artist to consider his genre to be just as important as his historical pictures; but he achieved this only by gradual stages. In style, these early pictures which Géricault painted of soldiers display something of the baroque composition and still more of the glittering, Rubens-like colouring of Gros, this being the only tradition in serious battle-painting to which he could link up.[2] It is,

[1] The latter probably being symbolic of the disastrous Russian campaign.
[2] Géricault, great artist that he was, did not disdain in the interests of his zealous research to copy, throughout his lifetime, military pictures of the preceding generation of artists,

however, characteristic of his artistic process that the picture of the *Officer of the Imperial Guard* (Plate 29a) painted as early as 1812, originated in the simple, visual impression of a horse rearing in front of a carriage which he had observed in the street. Side by side with his military pictures, Géricault made most concentrated studies after nature, in drawings and—a less usual practice —in pictures of horses and other animals, such as cats (Plate 28a), no less than of the human body; the muscles, the skeleton of the body, and even of the horse, received his closest attention. Such procedure was a sequel to, and a going beyond, the naturalistic studies after models as practised in David's atelier.

From 1816 to 1817, Géricault lived in Italy. Previously foreign painters going there had reacted, generally speaking, in one or other of two ways. If 'historical' painters, they were most impressed by the antique statues and the great Italian classicist and baroque artists. If inferior 'genre' painters, the animated outdoor life of the Italians proved their chief attraction. Géricault was perhaps the first painter to experience and combine both influences— and in this we recognize the artist whom we have so far followed. The Parthenon frieze, of which he saw a cast in Rome, Raphael, Michelangelo: these made the greatest impression on him. Hence he produced a certain number of mythological compositions, full of movement, in classicist and even baroque style. Yet the starting-point of his *chef-d'œuvre* in Italy, the work on which he concentrated all his energies, was—and this marks a significant stage in the history of painting—just a scene of contemporary life which caught his eye as a lover of violent movement, particularly in horses: the race of the wild, bare-backed horses at the Carnival in Rome. Even so, that he did not yet wish to produce a mere genre scene but still endeavoured to make of it an 'historical' picture denotes a phase in his development. The picture itself, which he never completed, is no longer extant, but a varied collection of sketches reveal his struggle to represent the theme. Individual motifs vary from a concrete rendering of the actual scene to an antique setting, while the compositions of the different sketches with their reminis-

---

which he considered dramatic: for example, *The Surrender of Madrid* by his master, Carle Vernet (copy in Bayonne), *The Revolt of Cairo* by Girodet, and *The Battle of Nazareth* by Gros (copy in Avignon).

cences of classical statuary, of Raphael, of Poussin are predominantly classicist even if occasionally intermingled with baroque-like traits (Plate 30b).[1] So that, although Géricault started from exact studies from nature, he still felt the necessity of monumentalizing, transforming his scene into a kind of antique one. This step is not great stylistically. Naturalism on the one hand and a vigorous classicism with its simplified and 'natural' poses and gestures on the other are intrinsically very close; the one style can easily be transposed into the other.[2]

After his return to Paris, Géricault still kept to this manner for a while, monumentalizing and classicizing, for instance, the scene of *Oxen driven to the Slaughterhouse* (Plate 28b). But just as he had been an actual eye-witness of this scene,[3] so it is symptomatic of his realist trend that he also studied, with intense application, the movements of wild beasts in the zoological gardens. At the same time, he went on with his representations of soldiers, showing the old Napoleonic army more and more from the angle of the unknown warrior, rendering passages of artillery, carts full of wounded soldiers (Plate 29b). But as the subjects increased in realism, the compositions, compared with the earlier representations of soldiers, became simplified, the baroque touch decreased; the tendency, already perceptible before he went to Italy, towards a more gloomy, nature-like sobriety in colouring, augmen-

[1] We reproduce here the sketch in the Walters Art Gallery, Baltimore, in which the spectators on the tribune are thrown on the canvas with a pictorial boldness not unlike that of Goya and Daumier, reminding one of Géricault's position between these two artists.

[2] The association between classicism and naturalism arising from the nature of the general outlook which lies behind the two styles is to be remarked not only in painting but in other branches of the visual arts as well. For example, the naturalistic bourgeois painting of the seventeenth century in Holland has its counterpart in a classicist architecture. Or again, the neo-palladian architecture in England from the twenties of the eighteenth century onwards was complemented by the naturalistic 'English' garden.

[3] Along the border of the Louvre drawing known as *Homme nu terrassant un taureau*, the actual driving of the oxen appears as in a frieze and it is self-evident that Géricault also created the main motif of the sketch as a result of watching the movements of oxen-drivers. Yet—characteristic of this early stage of his artistic career—the pose of the figure is obviously influenced by one of Michelangelo's *Bathing Soldiers* and perhaps the subject of the composition should ultimately be taken as one of Hercules' labours,

ted; his brush-work grew more robust.[1] Furthermore, a new technique furthered the new themes in his drive towards a democratization of art: Géricault was the first great painter to make a regular use of lithography—the technique of which he greatly improved—and in so doing he provided the less well-to-do with the products of his art.[2] Being himself well-to-do at this time, he did not attempt to sell pictures to individual patrons; in this way he was able to keep his artistic independence.

## IV

Political events in France were instrumental in approximating Géricault's art closer to realism. In 1816, a government frigate, the *Medusa*, was wrecked on a voyage to Africa. The ropes of a raft, intended to be towed by the life-boats, broke, the float drifted off, carrying 150 men. Tortured by hunger and thirst, fighting one with another, they were eventually sighted after twelve days; only fifteen were rescued, some in a dying condition. The disaster caused a great stir. The government was accused of having appointed as captain an inefficient favourite, and necessary means for safety had not been provided. The case was taken up by the liberal opposition and Géricault joined in wholeheartedly. Here was an instance of innocent people suffering, calculated to rouse both his political interest and his humanitarian instincts, while the dramatic aspects of the situation stirred his artistic sense—the turbulent sea, the lonely raft, despair, madness, dying people, corpses. He supplied lithographs for a pamphlet of accusation against the government, written by two survivors. But even more important, he decided to illustrate this contempor-

---

[1] An important forerunner of these 'unheroic' military compositions of Géricault is a picture by N. A. Taunay (1755-1830) which shows the French army—without Napoleon—crossing the Guadarrama pass in 1808 in a heavy snow-storm and mist (Versailles). Taunay himself derives from the realistic genre painting tradition of the eighteenth century.

[2] His friendship with the well-known painter and lithographer Charlet, who celebrated the Napoleonic army, dates from this period. Charlet's lithographs contributed greatly to the popularization of the Napoleonic legend—that important weapon in the hands of the liberal opposition. Géricault's closest friend, the painter Horace Vernet, the son of his first master, was an ardent supporter of the opposition and an admirer of Napoleon; his large circle, to which Géricault belonged, was entirely composed of liberals and of Napoleonic soldiers.

ary event in a large-scale picture. He chose for representation the moment when the rescue ship is hailed from the float, while around lie the apathetic, the dying and the dead (Plate 31a). When David was in opposition to the ruling Royalists, he gave vent to his feelings, as we have seen, in grand classical themes: in the *Horatii*, the *Brutus*. Later when he and other classicist painters used contemporary events, it was to represent outstanding achievements of the existing régime, whether of the Republic or Napoleon. Here a further stage has been reached when a contemporary event lacking a hero, concerned with nameless people, becomes the subject of a large picture and is painted as a direct means of opposition.[1] In fact, motives of opposition apart, it is as if, for instance, in Gros' picture of *Napoleon on the Battlefield of Eylau*, or that in which Napoleon is shown visiting the pestilence-stricken in Jaffa, the corpses and the sick had become the main figures and the Emperor been omitted. Gros, in the meantime, had yielded to the appeals of David who, exiled in Brussels as a regicide, out of contact with political reality and living in an old-fashioned mythological-classicist dream-world, wrote urging him to abandon his earlier naturalistic exploits and do his duty as David's deputy and leader of the classicist school in France. And it is worth noting that Gros' subservience to the Bourbons, whose retrograde tendencies made them, in an historical sense, 'out-of-date', propelled him in the same direction, that of an empty classicism. For the Government preferred antique, mostly harmless mythological, themes or religious ones or scenes showing the exploits of the Bourbons—all in a dry, classicist style.[2] So, while David and

[1] Still, opposition expressed in painting was not considered so serious as active participation in politics. Géricault even obtained a medal for his picture; having worked so hard on it, in the end he was rather distressed that it should have caused so much political controversy after its exhibition, and in fact he cherished a hope that the State would buy it. But there was, of course, no possibility of this, considering the 'opposition' subject. Instead, to show what kind of pictures were expected of him, the Government commissioned him to paint a religious picture (*Vierge du Sacré Cœur*). That he passed the order on to Delacroix rather than execute it himself, reveals, significantly enough, Géricault's mentality.

[2] Under the Restoration, the so-called school of Lyons came into its own: painters brought up in David's atelier and already known under Napoleon for their Bourbon sympathies depicted scenes of Royalist France in the past—Henry IV, Francis I, Joan of Arc—elaborating with accuracy the costume accessories. Ingres, an ardent supporter of the Restoration régime, in many pictures came near to this 'genre troubadour'.

Gros, both out of contact with reality and progressive thought, were losing their earlier *élan*, Gros, in particular, unable to find a way out of this hopeless artistic dilemma, even ending his life by suicide, it now became the turn of the young Géricault; having the necessary contacts he was well equipped to lead the way of modern art, doing for his generation what David and Gros had done for theirs.

Just as the whole incidence and theme of Géricault's *Medusa* picture mingled precedent with innovation, so also the method of its preparation and construction. Before Géricault's time, it was not uncommon for artists to make portrait studies for large pictures of contemporary events and attempt to render as accurately as possible the actual localities in which the events took place (David, Gros); but there had been nothing comparable with Géricault's most exact studies. He questioned the survivors minutely and portrayed each one of them with their features marked by intense suffering. He studied corpses and the sick in hospitals. He even had a model made by the same craftsman who built the raft and went to the coast to observe its movement afloat on the waves, studying, at the same time, atmospheric effects and the formation of clouds over the sea. It is really true to say that never before had an 'historical' picture been built up on the basis of such extended naturalistic research.

There is as much variety in these sketches as in those for the *Race of the Riderless Horses* (Plate 30b), but the classicism of the finished *Medusa* picture had, of necessity, to be more realistic, for the figures had of course to appear to be contemporary people. Within the realistic classicism of this painting,[1] with its crescendo from immobile to more and more active figures, all arranged in intersecting diagonals, certain moderate baroque elements of a residuary nature are also involved, as so often in the past with this kind of classicism.[2] If we wish to elucidate the relation of the *Medusa* to baroque and

[1] Indeed the picture was regarded by critics at the exhibition as being in the same line with those of the David followers.
[2] The term 'early baroque' which has lately been suggested for the style of the *Medusa* does small justice to the complexity of the picture. Baroque as the main term of definition is certainly misleading, as regards also the whole position of the *Medusa* in art-history. There can be no question of a baroque revival—whether early or high baroque—in Géricault, least of all in a painting by him representing a contemporary event.

classicist tradition and so determine its position in art-history, we must think primarily of the point where the two styles met for the first time and where classicism at its then highest possible peak, opened up for the first time a real possibility for a baroque style: that is Raphael's frescoes in the *Stanza d'Eliodoro*, so greatly admired by Géricault at the time he was painting *The Race of the Riderless Horses*. Whereas Correggio and Rubens exploited this potentiality to develop a style possessing many irrational elements, Caravaggio led the style of the *Stanza d'Eliodoro* toward a forceful, precise classicism, based on a pronounced naturalism without vagueness—even if moderate baroque elements do coexist in it, as in the late Raphael himself. Caravaggio laid more emphasis than did Raphael on vehemence of movement, on variety of expressions; he departed farther from a centralized composition; above all, his glaring light, strongly contrasting with deep shadow modelled the figures and stressed their solidity and reality, brought out clearly spatial relationships and held the composition together. All these elements were made use of by Géricault. Even during the second half of the eighteenth century, Caravaggio's art had greatly assisted Vien and the young David in their efforts to develop a naturalistic-classicism. And to Géricault at this given moment probably no artist of the past was so necessary as Caravaggio with his luminous classicism and naturalism, in its own day, so militantly progressive.[1] Besides the strictly classicist stream Raphael-Domenichino-Poussin-

[1] It is perhaps relevant to mention here that the artistic situation in Rome about 1600 was similar to that of about 1800 in Paris. The former contained the latter, as all later development of art for a long time to come, in nucleus. There were, in Rome about 1600 three main currents: the classicist-mannerist (e.g., Cavaliere d'Arpino), the classicist-baroque (e.g., Annibale Carracci) and the naturalist-classicist (Caravaggio). These three currents persist from that time; were present still in French painting two centuries later. Each French stream can be traced back to its predecessor of 1600: Ingres to the classicist-mannerist, Delacroix to the baroque, Géricault to the luminous, naturalistic classicism of Caravaggio. To what extent the similarities between the corresponding styles also cover similarities in outlook of the various social groups from which these styles emanate, could be shown if we were to examine the social and political conditions in Rome about 1600. As regards evolution itself, in a general way, the whole course of the three currents shifted with time, with the further development of the middle class, ever more to the 'left', that is to say, they all three became, each in its own way, radicalized through the infiltration of naturalism. But the 'radicalization' of the first two, intrinsically unnaturalistic, was, of necessity, only relative. The persistence of the more mannerist and

David, which we have already discussed, and which perhaps suffices to explain some of the sketches for *The Race of the Riderless Horses*, there is the other more 'leftish' one: the Raphael-Caravaggio-naturalist element in the young David-Géricault's *Medusa*. The *Medusa* with its restless, yet clearly organized composition, its individual motifs of self-contained value, undisturbed by any baroque twist, its bodies naturalistically modelled by means of light and shade, with emphasized expressions; the *Medusa*, then, marks the climax of this second artistic current. Furthermore, that most important factor in the logical construction of the *Medusa*, the increased accentuation of movement, already had its classic archetype in Raphael, and later in Caravaggio. On the long road from the sketches of the *Medusa* to the finished picture, Michelangelesque poses—poses often for their own sake—tend to disappear or are transposed, as already in Gros, into a more naturalistic vein and made to serve within a new, modern, rational content. Géricault, with the help of the whole naturalistic heritage of the past, including that of Gros, achieves a degree of progressive naturalism, far in advance of Gros. For the *Medusa* denotes not only the climax of an artistic current of the past in which the more consistently baroque elements had, in fact, been diminishing step by step from Caravaggio to Géricault, but implies, at the same time, its disintegration, thus opening the way for a more explicit naturalism in the immediate future.

V

Géricault, of course, sought out not only those forces of the past which best could help him towards naturalism, but still more so those of the present. These he could find only in England—the country with the most developed middle class and the most developed middle-class painting. Few French painters of any standing have gone, as Géricault did, first to Italy and afterwards

---

more baroque currents was of either a conservative or only slightly progressive nature, while the link-up with Caravaggio and the further realistic evolution of his style strengthened the most progressive middle-class current of the time, that is, Géricault. It is not surprising that Caravaggio, especially through his Spanish imitators, should later have played a prominent part in Courbet's development likewise.

to England.[1] Yet the chronology of Géricault's two equally important journeys expresses symbolically the whole development of his art. In England (1820–21) he did not utilize scenes of contemporary life, as in Italy, merely as a starting point for classical scenes; he now regarded them as ends in themselves. Nor did he choose, as recently, some spectacular contemporary event but preferred everyday occurrences such as races, street or stable scenes. Géricault's recorded remarks in England show to what extent his impressions there impelled him in the direction of naturalism. He says at first somewhat scornfully that the only good painting to be seen is landscape and genre, that historical painting scarcely exists. Yet it was, in fact, precisely the former which was in conformity with his own development. He was quickly captivated by the colourful English landscape and genre painting—the nineteenth-century results of Dutch bourgeois painting of the seventeenth century—especially by Constable, but also by Morland, and the sketches of the young Wilkie.[2] For, in contrast to Italy, he saw only contemporary art in England. And the whole English life with the singular part played in it by his favoured horses and races and the artistic expression of these in sporting engravings and pictures, were naturally most congenial to him.

---

[1] Corot certainly did so, but his English journey, very late in life, assumed far less importance, than Géricault's. The same chronology applies also to Stendhal and with him it had more significance. It is probably no coincidence that such a close resemblance exists between the political, literary and artistic ideas of Géricault and Stendhal.

[2] Géricault was not the first Frenchman whose already pronounced naturalism became enhanced through his stay in England and direct contact with the similar tradition of English art. Loutherbourg, who came to England in 1771, settling down for the rest of his life, underwent the same experience and, in his turn, influenced English painting (and the stage) in a realistic sense. (Cp. E. Wind, 'The Revolution of History Painting', *Journal of the Warburg Institute*, II, 1938.) A striking picture by him showing a Methodist divine preaching in the open (1777, National Gallery of Canada, Ottawa; Plate 149b, *Hogarth and his Place in European Art*, F. Antal, 1962) is undoubtedly affected by Hogarth's art and he was perhaps the first to represent, in a picture, workers road-digging (London, Messrs. Agnew). His paintings of sea-battles are the immediate forerunners of those of the early Turner. The reciprocal influence of French and English art in the matter of naturalism could also be achieved through engravings, which were usually more realistic than pictures. Debucourt in his famous colour-print, the *Promenade de la Galerie du Palais Royal* (1787), was certainly influenced by Rowlandson's Vauxhall print and he, in his turn, left his impression on English graphic art.

In his famous *Epsom Derby* (Louvre), he records merely the momentary visual impression of a race he attended. Four racing horses in action, with their jockeys—the only other object being a white marking post. This picture is usually regarded in the literature of art-history as a sudden miracle, lacking antecedents. The reason is, no doubt, that the art-historian, however unintentionally, regards even Géricault as still a 'history' painter, and largely ignores the lesser types existing already before his day (genre pictures, sporting pictures, landscapes, engravings), that is, works of art not conforming to the grand manner, or actually answering to the taste of a lower social stratum, and which are deemed not worthy of study to the same extent as 'historical' pictures. Hence, while art-history has registered the increasingly naturalistic aspect of the traditionally well-known 'historical' pictures, it has tended to overlook the further fact in the development of naturalism, that the lower genres, intrinsically more naturalistic, and which make up by far the greater proportion of the art of their period, were slowly, step by step, rising to higher rank.[1] It is easy to find resemblances to the *Epsom Derby* in previous English racing prints,[2] as regards the theme, the composition as a whole and the individual motifs, even that of galloping horses with all four legs outstretched. As a rule it is only in engravings after English racing pictures, not in the pictures themselves, that one finds close similarities to Géricault, for this type of representation is far surpassed by the *Epsom Derby* in pictorial treatment. Géricault's work is no longer a mere dryly painted sporting picture—as frequently adapted for the purpose of engravings designed for lovers of sport; it is painted at the artist's highest level, when he was applying in his own way a well-assimilated understanding of the colour intensity, the freshness, the broad and nervous brushwork of the best English painting. This English 'influence' reassociated Géricault with the French eighteenth-century pictorial tradition which had been temporarily abandoned at the end of that century in favour of a classicism that promised, despite its smooth-

---

[1] It would be too dogmatic to suggest, for instance, that engravings are first elevated to the rank of genre pictures, and later, genre pictures to that of 'historical' pictures; but, as a generalization, one may say that a development of this kind, even if in a spasmodic way, does take place in course of time.

[2] In fact, Géricault's first master, Carle Vernet, had already painted horse races, obviously influenced by English sporting engravings.

ness, to be more naturalistic. Through this linking up with both the French past and the English present, Géricault may truly be considered a forerunner of Impressionism. But even in this most naturalistic picture, *Epsom Derby*, one can detect Géricault's classicist roots: the bareness of the composition and its simple lines betray them. (There is here no question any more of anything baroque.)

Even in the most 'advanced' works which Géricault did in England, whether oil-sketches of races or water-colours or pen drawings of circus scenes, one can feel this classicist basis.[1] The most amazing sketch of all, that of *Four Jockeys* (Plate 30a) in the Museum of Bayonne, where the impression of instantaneity is even more effective than in the *Derby*, already foreshadows Degas and Manet. The step from the various Poussin-like classicistic races of bare-backed horses to this almost impressionist, yet solidly compact, and firmly moulded sketch, achieved during a short period of three or four years, seems, at first sight, almost incredible. Indeed, we get here not only the now familiar impression of a transition from classicism to naturalism but also—as so often in Géricault's late phase—an early foretaste of that phenomenon often met in French figure compositions of the second half of the nineteenth century, the thin borderline between classicism on the one hand and the different shades of Impressionism on the other.[2] When Géricault made a monumental water-colour of *Ploughing* in England (New York, Sterner Gallery)—a novel theme for French (if not for English) art—classicism and realism were fused more organically than later in the more primitive simplifications of Millet, yet with a greater emphasis upon the realistic. Despite its quiet lines, the whole drawing, with its sparkling patches of light and shade, is vibrant with life: the workers, the ploughs, the horses, the landscape. The lithographs, produced in London, are fundamentally similar in style. Géricault now continued to work in this new democratic technique—as yet little practised in England—on a much larger scale than in Paris. He depicted (*Various Subjects drawn from Life and on Stone* by T. Géricault, London, 1821) in addi-

---

[1] That Géricault went as late as 1820 to Brussels to pay his respects to the exiled David shows how greatly he esteemed the latter's art. He must have felt strongly even then that his own art was ultimately a continuation of David's.

[2] E.g. Manet's *Déjeuner sur l'herbe*, Degas' paintings in the sixties and early seventies, Pissarro's peasant pictures, Seurat's *La baignade*.

tion to stable and sporting incidents (e.g., *Smithy*, *Boxing Match*), many street scenes of all kinds which show his keen understanding of London life, particularly how interested he was, agreeably with his extreme liberal outlook, in the sad and dreary aspects of the popular, industrial quarters.[1] He shows a street beggar, fainting from hunger in front of a bakery, a blind piper playing in an empty street and above all every kind of labour (Plate 31b), whenever possible connected with horses and carts. Big coal waggons, which he had already practised upon in Paris, now became a favourite theme. As with all manner of working horses, he recorded the individual features of these heavy, shaggy beasts, evidence here too, of his 'democratizing' purpose, for they represent the other extreme from refined racing animals or the artificially slim ones of Carle Vernet.

Perhaps no one in England has so genuinely carried on the art of Hogarth, the most unambiguously and militantly bourgeois-minded of all English artists, as this French cavalier coming to these islands imbued with humanitarian ideas and giving expression to them in his London lithographs. In comparing the general outlook of the two, it is evident that Géricault, living in an age when a more pronouncedly liberal thought was possible, sympathized with the poor in a more positive manner than did Hogarth, with his all-round love of caricature. In fact, most of Hogarth's achievements were carried a step further by Géricault, his representation of new, more genuine, unvarnished everyday scenes and happenings; his elevation to the sphere of paintings, of subjects previously treated only in engravings; his more vigorous and realistic mode of painting and the diffusion of his work by means of graphic art. But while Hogarth had still to be, though at heart unwillingly, a partisan of historical pictures—a necessary condition, as he believed, of raising the social standard of the artist—a hundred years later this struggle was no longer necessary and so the whole sham of historical painting became superfluous.[2] This levelling of the different ranks of painting was mainly carried through by Géricault himself and it was precisely his stay in England

[1] A comparison with Wheatley's *Cries of London* shows a revolution in the themes of street scenes and, allied to this, a more realistic formal idiom.

[2] Diderot's suggestion that every picture in which human beings were represented should be called historical (and no longer genre) marks an important transitional stage in the development of this democratization of art.

which finally convinced him of the correctness of this point of view. Decisive was the change from his initial disappointment at the lack of historical painting in England to the opinion soon to be expressed in a letter from this country: 'J'abdique le cothurne . . . pour me renfermer dans l'écurie.'

After his return from England Géricault had but two years to live (1822-24). A great part of this time he was obliged to spend in bed, the consequence of a fall from his horse which aggravated a previous malady. Although he was thus only able to do small pictures, sketches and lithographs, yet the outstanding characteristics of his art became greatly intensified in the tragically short space of time at his disposal. Innumerable subjects now turn up, then vanish again with lightning rapidity, all of them treated with a realism more novel and consistent than ever before. We can touch upon only a few of the themes here.

During the closing days of his life, he was as interested in politics as ever. In his persistence to portray topical events with a propagandist purpose, he is perhaps the first great artist whom one can call 'painter of actualities'.[1] He who once had represented in lithographs battle-scenes from even the little-known struggle of Chile for its liberty, as a matter of course now shared the general enthusiasm for the Greek war of Independence and made sketches of individual scenes in it. He took every occasion in his drawings and lithographs to make a stand against the French government's reactionary home and foreign policy, evoking as argument—like Béranger in poetry—the relative liberalism of the Napoleonic period. He even played his part in creating the Napoleonic legend by depicting the Emperor on several occasions on the battlefield—a subject he had almost entirely avoided during the latter's régime. When the government of Louis XVIII intervened in Spain on behalf of the absolutist monarchy and clerical reaction, Géricault showed how, in contrast, Napoleon's government had opened the gates of the Inquisition. Among oppressed races, the negroes especially attracted him. Even in the *Medusa*, the topmost figure, feverishly waving a piece of cloth, is intentionally a negro. Joining the liberal campaign against the slave trade (which was to have positive results only in '48), he now planned a large picture exposing the cruel treatment of negroes. This was never carried beyond the prepara-

[1] Cp. Duc de Trévise, 'Géricault, peintre d'actualités', *Revue de l'Art ancien et moderne*, 1924.

tory drawing (Plate 32b) (Paris, École des Beaux-Arts) which shows that the composition, while retaining a certain classicist structure, would yet have been much simpler and more effective than the *Medusa*.

Géricault's delight in depicting negroes is doubtlessly associated with the general romantic liking for the exotic—we have seen a similar phenomenon in the much more conservative Girodet. But Géricault's interest probed deeper than this; just as his predilection for the race had a real political basis, so his artistic expression of it had a concrete, naturalistic one. Through his customary intensive studies, he was able to produce realistic representations of negroes as none before him. Just as concrete, even scientific, was his preoccupation with the 'horrific' about which romantics were enthusiastic in a vaguer way. He painted—according to a prescribed plan—a series of lunatics, both men and women, patients of the Salpêtrière, to illustrate a book which the chief doctor of the hospital intended to write on mental diseases; here lies the whole difference. Géricault painted the pathological subjects stripped of any 'romantic' exaggerations and with careful observation of distinctions among individual abnormalities.[1] He portrayed, among others, a monomaniac who supposes himself to be a military commander and wears his inmate's number as a medal (Reinhardt Coll., Winterthur)—significant this of a painter who likes soldiers but dislikes, as a liberal, overbearing militarism. The immense artistic resources brought into play to render these different types of insane, make Géricault's pictures astonishing for so early a date as 1822. A sketchy one of a mad young woman (previously Eisler Coll., Vienna) which shows a close acquaintance with English portrait painting but pursues the possibilities of it much further, comes decidedly close to works of Manet. When Géricault made a self-portrait during these years (Rouen), he applied the same searching naturalism and put to a most merciless use his supreme mastery of the means of pictorial expression in recording the ruined constitution of a man, still young in years.

---

[1] Géricault even painted (at the time of the *Medusa*) heads of dead men he obtained from hospitals. The horror of the last moment is obviously expressed in such heads and it is no accident that Victor Hugo was fascinated by them, taking them, as tradition too has it, to be those of men guillotined. But Géricault was concerned as always with the impression of nature just as much as with the shock of horror: naturalism once more extended to 'romantic' subjects.

It was in this last phase that he painted his own derelict limehouse outside Paris (Plate 33a) (Louvre)—the object of an ill-fated speculation—when on a visit there he suddenly recognized its pictorial aspects. So Géricault treated a very simple motif, similar to that of ploughing in the country, no longer as a mere water-colour, but raised it to the subject of a painting. Rugged cart horses, some standing beside an empty waggon, others drawing a loaded one, a muddy road, are the other features of the picture, the few figures ranking on the same plane, with nothing 'poetic' about them as in historical or mythological landscapes. The derelict, dusty corner of nature is rendered with genuine sincerity and an economical distribution of light and shade, unknown even to the best so-called 'genre' painters of the period.[1]

More so than Ingres, ultra-conservative in outlook[2] and on the whole in his art, more so even than the mildly conservative Delacroix, Géricault was forerunner of most that was progressive in nineteenth-century French painting, especially in the immediately succeeding decades. Millet, who was still to come, is almost surpassed here and the way open for Courbet, that enthusiastic admirer of Géricault. At a time when advanced middle-class painting aimed increasingly at a thorough conquest of nature, it was Géricault who made possible the decisive transition from a naturalistic classicism to a more consistent naturalism, and to such an extent that certain elements in him actually foreshadowed Impressionism:[3] it is this which determines his true place in art-history. In line with the whole previous organic development, and carrying it, step by step, from one logical sequence to another, Géricault became—and in this there is no contradiction—a revolutionary master. The

---

[1] An English picture is, in fact, its closest forerunner: G. Garrard's (1760-1826) *Brewery Yard* (1784?, Whitbread Coll., Biggleswade)—a far more solid and realistic work than any of Morland's. Garrard, who portrayed agricultural shows and published engravings of such subjects as *Improved British Cattle* was a scientifically-minded painter.

[2] He thoroughly disapproved of Géricault's *Medusa*, apparently because it appeared to him in every sense too topical, and wished it to be transferred to the Admiralty, just as the two large early pictures of soldiers to the War Office, 'pour qu'ils ne corrompent pas le goût du public qu'il faut accoutumer à ce qui est beau. . . . Je ne veux pas de cette Méduse et de ces talents d'amphithéâtre qui ne nous montrent de l'homme que le cadavre, qui ne reproduisent que le laid, le hideux.'

[3] The important relation between Géricault and Daumier in content as well as in form should at least be noted.

intensity of this short career, wrongly termed 'romantic', is almost without precedent. Although Géricault himself considered his works of a merely preparatory character, there is perhaps no other example in art-history, Giorgione alone excepted, of a painter who died as early as thirty-three having yet created an art so far-reaching in its effect and having produced a change so profound in the actual artistic situation.

# 2

# *The problem of Mannerism in the Netherlands*

Research into Northern mannerism, especially in the Netherlands, at the end of the sixteenth century, is in full swing. Books on Bloemaert by Delbanco, and on Wtewael by Lindeman have appeared recently, as well as a long treatise by Steinbart on Frederick Sustris and Pieter Candid, whilst works on Spranger, van Mander, Cornelis van Haarlem and Heintz are, according to report, in preparation. As papers on Goltzius, Hans von Aachen, Rottenhammer (as well as an older one on Spranger) have been available for some time, we are definitely advancing towards an elucidation of the whole material.

The book on Bloemaert[1] presents above all a useful and critical, although not quite complete, catalogue of the artist's works. Several hitherto misleading paintings have rightly been excluded from the *œuvre*.[2] Bloemaert's development, which can easily be deduced from his paintings, and which is so characteristic of the whole trend of Netherlandish art, is at last perfectly clear from the chronologically arranged *œuvre-catalogue*. The artist began as an extreme mannerist; the mannerism was later moderated by the impact of classicism; and this, in turn, was followed by a Caravaggesque period, likewise toned down by classicism. The unvaryingness of the scheme of

---

[1] Gustav Delbanco, *Der Maler Abraham Bloemaert*, Strasbourg, 1928.

[2] Although Delbanco is unable to suggest attributions for the most important amongst them. This can however be done in several instances: the painting of *Venus and Amor* (Coll. Nostitz) signed *A.B.* is by Antonis Blocklandt (attribution by Wescher). *The Raising of Lazarus* (Munich) and probably also *Hercules and Omphale* (Copenhagen) are by Abraham Janssens (attributions by Longhi and Bauch). According to Burchard, the Munich picture is identical with the painting which already Decamps mentioned as Janssen's *chef-d'œuvre* and which at that time was in the possession of the Count Palatine at Düsseldorf.

composition—basically still dependent on mannerism—made it possible for the late works of this long-lived artist (he died in 1651) to form a bridge to the italianizing style of Poelenburgh, Berchem and the Academists. Bloemaert's *œuvre* demonstrates that mannerism was never completely ousted by the so-called 'great' period of Dutch painting. How short-lived that period actually was is manifest from the fact that the beginning and end of Bloemaert's working life span both mannerism and the new Italianizing style.

The way in which Delbanco extricates himself from the problem of the origin of Bloemaert's style is typical of the existing state of art-historical research. Whenever one consults a book or article by a German art-historian in which the author attempts to indicate the foreign sources of Netherlandish painting of the late sixteenth century, the same picture presents itself. To those who have not themselves studied the history of Italian painting of the period, only two sources of help are available: first, the Fontainebleau theory, put forward by Dimier in 1905, adopted by Kauffmann in 1924 and, though unfortunately wrong, thereafter accepted as cut and dried; secondly, the corpus of illustrations of Florentine and Roman works (all that were obtainable in 1919) contained in Hermann Voss's book on late Renaissance painting (*Die Malerei der Spätrenaissance in Rom and Florenz*, 2 vols., Berlin, 1920), from among which one can pick and choose stylistic parallels according to one's fancy. Delbanco skirts gingerly around the disquieting Fontainebleau theory and eventually sees that the facts compel rejection, both in principle and in detail. Like Hirschmann before him, Delbanco finally comes to the conclusion that it was Spranger's style—originating in Italy and introduced to Haarlem around 1583/85 by van Mander and Goltzius—that was decisive for early Dutch mannerism, and so for Bloemaert too. Delbanco indicates Correggio and Parmigianino in general as the main Italian influences on Dutch mannerism, and among the illustrations to be found in Voss's book points particularly to Raffaellino da Reggio's *Tobias and the Angel* (Borghese Gallery, Rome), because of its general resemblance to works by Spranger and to Dutch mannerism; in this resting on the authority of van Mander, who discussed Raffaellino at length and praised him highly.

From the confusing way in which Lindeman[1] assembles and presents his

---

[1] C. M. A. A. Lindeman, *Joachim Antonisz Wtewael*, Utrecht, 1929.

material on Wtewael, nothing can be learned of the artist's development. He treats successively Wtewael's signed and dated paintings; then the signed but undated; then the unsigned and undated; and lastly the drawings. From all this elaboration nothing positive emerges, for the sufficient reason that the author has no idea of the overall development. In fact the story is not complicated: Wtewael's paintings vary but little; only the early works, which did not yet conform to the later settled manner of composition, are of surprising intensity. If one knows how to integrate these works—for example *The Flood* (Germanisches Nationalmuseum, Nuremberg (Plate 12a))—in correct chronological order into the general evolution of Dutch art, the whole problem of the artist's development is solved. Moreover, Lindeman's catalogue of the drawings is uncritical from the point of view of connoisseurship: works by Bloemaert, copies, and so on, are included. Lindeman rejects the Fontainebleau thesis on the ground that the style of Fontainebleau derived from Parmigianino, whereas Dutch mannerism depended on Michelangelo and his followers in Florence and Rome.[1] Taking together all relevant statements from hundreds of pages of 'basic'—all-too-basic—considerations, the following bare names emerge from among Roman and Florentine artists as parallels to Wtewael: G. B. Pozzi, Michelangelo Carducci, Federico Zuccaro, Zucchi,[2] Marco Pino; and as parallels to Bloemaert: Federico Zuccaro, Roncalli, Poppi, Poccetti. It appears—although one cannot be absolutely sure—that Lindeman regards all these as followers of Michelangelo. However, one can be certain of one thing: this Tusco-Roman haphazard assortment has resulted from flicking unsystematically

[1] The style that prevailed in Tuscany, Rome and Holland at the end of the century, appears to Lindeman to be 'Michelangelesque'. He applies the term 'mannerist' only to the followers of Michelangelo, whilst Parmigianino, El Greco, etc., are for him not mannerists at all. It is impossible to go into this and similar naive arguments, extending often to more than a hundred pages, on mannerism in general, on mannerism in Italy, on the relationship between Italian and Netherlandish art, etc. They constitute a heroic, if unconscious, 'toil of Penelope' *vis-à-vis* the results of contemporary scholarship.

[2] Voss, although only in casual hints, was the first to point to Zucchi, Boscoli, and Salimbeni, the three Central-Italian artists who were to prove of such importance for the North. However, in order to discover these three references, one would have to read through all two volumes. Lindeman found only the reference to Zucchi, mentions him in a sentence, but does not follow up this highly important trail.

through the illustrations to Voss's book. Where Lindeman deals with individual works in detail and tries to enumerate the surprisingly few concrete, stylistic connections upon which he has stumbled, almost his sole and well-worn text-book example is the Venetian, Jacopo Bassano. Only once, apropos of *The Baptism* (Copenhagen), is Lindeman reminded of specific central-Italian mannerist works, namely, two drawings, illustrated by Voss: Federico Zuccaro's sketch for his Florentine stage-design, and Barocci's preliminary drawing for the painting *Moses and the Burning Bush* (Uffizi).[1] Accepting them without criticism for the moment, these constitute the entire positive achievement of this piece of secondhand art-history.

The paper by Steinbart on Sustris and Pieter Candid[2] is a welcome addition to our knowledge of these two artists, which, apart from the out-dated book by Rée, has hitherto been based on Basserman-Jordan. Steinbart makes some corrections and publishes several new or little-known works and, particularly, many details: frescoes by Sustris in the Fugger House at Augsburg, Schloss Trausnitz, and the newly discovered ones by Candid in the old palace at Schleissheim, as well as numerous studies by both for these and other works, engravings after them, and so forth. With these two Netherlanders, who worked at the Bavarian court and who constitute a perfect parallel to the Dutch mannerists around 1600, it is so obvious that their style originated in Florence that there can be no question. Their careers alone provide proof of this, for both Sustris and Candid collaborated, under the personal supervision of Vasari, on the paintings and other artistic activities which the latter carried out in Florence (and Candid performed this same function in Rome). In addition to Vasari, Steinbart suggests as other likely influences, Gherardi, Taddeo Zuccaro, Barocci and Bassano.

[1] Both references are typical of the dependence on the illustrations in Voss's book. Lindeman of course knows only the London fragment of the drawing for the Florentine theatre-décor that Voss discovered and published, but not the complete design—four times as big—in the Uffizi, which Voss also mentions, but unfortunately does not reproduce. As for the Barocci drawing a reproduction of the finished fresco is to be found in W. Friedländer's *Casino Pius IV*. Lindeman connects Bloemaert's *Raising of Lazarus* with F. Zuccaro's fresco of the same subject 'in the Palazzo Grimani'. The fresco is in the Cappella Grimani in S. Francesco della Vigna, Venice.
[2] Kurt Steinbart, 'Die niederländischen Hofmaler der bairischen Herzöge' in *Marburger Jahrbuch*, IV, 1928.

# The problem of Mannerism in the Netherlands

A correct and more or less definite solution to the problems of style and development of the two artists of whom Delbanco and Lindeman treat, and which in Steinbart's work is resolved almost naturally, can be found only if the whole development of both Netherlandish and Italian painting, as well as that of the School of Fontainebleau, is examined—best of all each from within its own tradition. Actual 'influence' by the art of one country on that of another is possible only where both, each having set out from its own premises, have taken parallel paths, and where, moreover, the manifestations in the one country are to some degree in advance of that of the other. Naturally such relationships do not necessarily occur only between contemporary styles; quite as significant as the revival of stylistic analogies in earlier periods of the art of one's own country may be the reversion to an earlier phase in the art of another. Once these internal analogies are clear, then the external 'similarities' (influences, adoption of motifs) between the arts of the various countries stand revealed, in surprising numbers, and almost spontaneously, so that there is no need for a spasmodic, haphazard search for occasional appropriations of detail.

On no account do I wish to imply that I am in a position to satisfy all the demands just stated. I know only too well from personal experience, what difficulties, primarily technical, have to be overcome; I can do no more than advance a little the programme sketched out above. Beyond mere criticism, I would wish, so far as it is possible at the present time, to say something positive about the relationship between Netherlandish and Italian mannerism and, disregarding the strict difference between 'parallels' and 'influences', to offer a few concrete findings.[1]

Netherlandish art of the sixteenth century was of course not the outcome of passive submission to Italian influence. Northern artists themselves took active steps to bring about the assimilation and at some cost, furthering it by travel and by study of engravings. They sought it out because it was already

---

[1] The examples have been selected at random. They could be extended almost indefinitely and could probably be substituted by more telling ones, once the whole material, especially the Netherlandish, has been thoroughly explored. What mattered most was that the examples should be on the right lines. Some of the comments go, strictly speaking, beyond the immediate scope of this paper but because of their wider interest I included them.

in their own vein. It was not a case of straightforward adoption or imitation of the type of figures and compositions which had been created in Italy.[1] It was a process rather of assimilation: an absorption of the essence, adapting it to the native Northern manner, then developing it in the general sense of mannerism, the one creative stylistic force of the sixteenth century. By the exertions of the Netherlandish artists, the tremendous lead Italy had gained was evened up by the end of the century. Parallels between the two countries, which at the beginning of the century had often been of a merely general nature, became, during its course, more and more marked and clearly discernible. Stylistic characteristics of the late Middle Ages were naturally more pronounced, and appear more purely late-gothic in Northern than in Italian painting at the beginning of the sixteenth century. But now, in spite of—or indeed because of—the fact that the North fought with weapons of the Italian Renaissance against its own gothic inheritance, the 'mediaeval-Northern' aspect of Netherlandish painting was intensified; not only the achievements of the Renaissance, but those also of mannerism, were in process of absorption from Italy. Moreover, Italian mannerism, which in many ways represented a continuation of the Quattrocento gothic style itself, had a strongly 'mediaeval-Northern' basis,[2] so that Italy is to be seen not only as a giving, but also—and most emphatically—as a receiving, partner of the North. This is one of the fundamental hypotheses for the following discussion. In Italian art, during the course of the sixteenth century, continuous and very pronounced classical and baroque tendencies were discernible, again and again to be overwhelmed by those of mannerism. In the North, this intermediary stage of an occasionally emergent classical style occurred only under certain conditions and in a limited sense, and the tendency towards mannerism was much more widespread and intense. It is therefore

[1] M. Dvořák, 'Über die geschichtlichen Voraussetzungen des niederländischen Romanismus' in *Kunstgeschichte als Geistesgeschichte*, Munich, 1924.

[2] I have defined what I imply by 'latent-Northern' in my 'Studien zur Quattrocentogotik' in *Jahrbuch der Preussischen Kunstsammlungen*, 1924 and 'Gedanken zur Entwicklung der Trecento-und Quattrocentomalerei in Siena und Florenz' in *Jahrbuch für Kunstwissenschaft*, 1924/25. There it is clarified by means of descriptions and characterizations of the works. How far it is possible to operate in this respect with clear cut definitions, I hope to show in a review of the highly professional book by Th. Hetzer, *Das deutsche Element in der italienischen Malerei des XVI. Jahrhunderts*, Berlin, 1929.

more difficult in the North than in the case of the South to draw a definite line between what is 'still' pre-Renaissance late gothic, and what is 'already' post-Renaissance mannerism.[1] By accepting a style so akin to its own as that prevalent in Italy, the North checked straightforward development of its own native gothic tradition, thereby enabling gothic tendencies latent in the Italian style to flower the more strongly. Some Netherlandish artists—Heemskerck for example—seem particularly 'mannerist', just because they were inherently 'late-gothic' and 'retrograde'; being more mannerist than the contemporary Italian artists, they thereby even anticipated the general development. The unfolding of the arts in the Netherlands during the sixteenth century is so very complex because, in spite of the above-mentioned parallel development between North and South, each fresh impulse of style in Italy was only truly comprehended by the following Netherlandish one, which moreover at the same time superseded it. At the beginning of the century the influence of preceding generations of Italian artists still played a decisive part; during the course of the century, however, a craving grew for an impact of much greater immediacy and power. At first the Netherlandish artists took up the Renaissance style and transformed it into their kind of mannerism; later they took up Italian mannerism and again made Northern mannerism of it. The distance separating those who influenced from those who were influenced steadily diminished, until around 1600 the border-line disappeared. An elementary test is that whereas at the beginning of the century it would have been impossible to mistake an Italian for a Netherlandish picture, by the end of the century there existed numerous border-line cases in which it is difficult to distinguish Italian from Netherlandish art. Despite the undeniably distinctive character of its development, the fate of Netherlandish art in the sixteenth century was predetermined in Italy. It would seem necessary therefore to discuss first of all the evolution of Italian mannerism.[2]

[1] Cf. Breu and Filippino (*Zeitschrift für bildende Kunst*, 1928/29).
[2] I have therefore thought it proper to illustrate mainly unpublished Italian paintings that were of decisive importance for the North. Among the Italian works I have selected mainly Roman paintings because the possibly even more telling Florentine ones are in my book on Florentine painting of the sixteenth century.

## II

In my review of Longhi's paper on Italian-Spanish relations around 1600,[1] I attempted to isolate the Tusco-Roman mannerist style as one of the chief components of European mannerism generally. I indicated very summarily how this particular style, especially as it appeared during the last third of the century, decisively influenced the younger Netherlandish mannerism with which we are mainly concerned here. I also showed that this uniformly irrational style resembled in many respects the earlier mannerism about 1520/25. The close traditional connection between early Tuscan mannerism and the native, as it were pre-mannerist, gothic of the Quattrocento, had a great deal to do with the recrudescence of mannerism in this region during the second half of the century, and its concomitant revival of many late-gothic elements. A specific Northern streak had of course been inherent in Tuscan art for a long time, and it is typical that, for example, Zucchi, the representative of the later mannerism, borrowed from Dürer in the same way that Pontormo had done fifty years earlier, and before him Robetta had borrowed from Schongauer, and the members of Pollaiuolo's studio from the Master E.S. It is precisely this characteristic receptiveness to Northern art which explains the importance of Florence for later Northern mannerism.

The manifold compositional variations of early Tusco-Roman mannerism, mostly of a pronounced two-dimensional character, arranged usually in tier-formation, were more or less clearly demarcated from any North-Italian or other baroque tendencies. These were the characteristics towards which the whole development at that time was striving, and which the later international mannerism made its own and applied almost as a formula: decentralized compositions, emphasizing and exploiting the dynamic character of diagonals, and derationalizing the spatial element by means of complicated and playful exaggeration, as in Pontormo's *Martyrdom of St. Maurice and the Ten Thousand* (Pitti) (Plate 7a); Caraglio's engraving after Rosso's *Hercules and Achelous* (Plate 3a); Giulio Romano's frescoes in the Villa Lante; or Beccafumi's typically Sienese narrative style, heightened by contrasts of colour and chiaroscuro, e.g. the ceiling decorations of the

---

[1] In *Kritische Berichte zur kunstgeschichtlichen Literatur*, 1928/29.

Palazzo Pubblico in Siena. The figures became more or less detached from the *rationale* of the action depicted, whilst their formal qualities and spiritual significance were accentuated. Such stylistic characteristics were of course unthinkable without Michelangelo's Sistine ceiling and *Last Judgment*.[1] Important for the development of mannerism, as it has been outlined here, were the idiosyncratic but consistent styles of Rosso and Bandinelli. I refer particularly to Rosso's frescoes of the *Creation* and the *Fall of Eve* in S. Maria della Pace in Rome, and to Caraglio's engraving of the *Rape of the Sabines;* also to Bandinelli's *Martyrdom of St. Lawrence* (engraved by Marcantonio) and the *Massacre of the Innocents* (engraved by Marco Dente—in fact a mannerist interpretation of the engraving by Marcantonio of the same subject after Raphael). Some of these crowded scenes, frequently composed of figures in artificial poses which bear no relation to the subjects depicted, came very close to what the Northern mannerists were about to attempt, and were therefore of special significance for them.[2] The emancipation of the figure was destined to be of fundamental importance for a style, as arbitrary and disdainful of the merely sensuous and obvious, as that which was to prevail in Europe throughout the century. Also the subject-matter that was to be so characteristic of later international mannerism is already discernible: mythological or allegorical representations with nude, and particularly semi-nude, figures in whom crude force and erotic elegance mingle, the former emphasized by wild, fantastic gestures and grimaces, and by grisly, fabulous beasts, as in Rosso's *Allegory of Wrath* and the *Labours of Hercules* (both engraved by Caraglio)[3]; the latter by jewellery and other accessories cun-

---

[1] The unique and extremely complicated position of Michelangelo between classical art baroque and mannerism can of course not even be hinted at within the framework of this paper.

[2] *The Massacre of the Innocents* by Goltzius and Cornelis van Haarlem (even that by Floris, engraved by Ph. Galle) should be examined for their reminiscences of and borrowings from Bandinelli. Both for the North and the South, the origin of the mannerist exaggeration was the engraving by Marcantonio after Raphael (B.18).

[3] To enumerate all the Netherlandish adaptations and copies which derived from Caraglio's engravings—especially the *Labours of Hercules*, Rosso's *Gods*, and the *Loves of the Gods* of Rosso and del Vaga—would be like listing every instance of the word '*et*' in Livy. Hitherto nobody has even hinted that these relationships existed. Some interesting examples will be mentioned later in the text. In particular Goltzius's compositions (as they

ningly placed about the naked body, or preciously draped, as in del Vaga's and Rosso's scenes of the *Loves of the Gods*; or in Rosso's single figures of deities such as his *Venus* and *Mars* (engraved also by Caraglio). Further, in Giulio Romano's *Birth of Venus* (Sala delle Prospettive, Farnesina)—a transposition of Raphael's *Galatea*—we already find the sea-gods furnished with a profusion of those accessories—shells, fish, garlands of fruit, etc.—which were to be typical of later international mannerism. The predilection of Tuscan artists, such as Pontormo, Rosso, Beccafumi, for hard, angular, jagged, sprawling forms, combined with a pleasing, slender, late-gothic grace, was of special importance as part of that rhythmic interpretation of the human form, which was common to all the artists of the earlier mannerism and which appeared at its most pronounced in Parmigianino. Thus one frequently finds swelling and diminishing, or nervously vibrating flexible contours, which already corresponded to the style of Spranger and Goltzius.[1]

By the mid-century there prevailed in Central Italy, especially in Rome, a sculptural, academic, Michelangelesque style which was far less intensely mannered than that of the earlier mannerism around 1520/25. At the same time there swept over Northern Italy a style, influenced by Parmigianino, which developed especially the rhythmic and ornamental side of mannerism, and became of great significance for Northern Europe. In this connection one should consider particularly Schiavone's etchings[2] and those by the Veronese artist Battista del Moro; also the later engravings of Battista Franco, the second period of Jacopo Bassano and, to a certain extent, even the elegant, magnificent early style of Tintoretto. Closely related to this Parmigia-

---

[1] To cite only one example: compare Beccafumi's large figure of the Executioner in the *Decapitation of Spurius Cassius* in the Palazzo Pubblico at Siena with Goltzius's *Mars* (engraved by Matham). The prototype for both figures, by the way, was Rosso's figure of *Mars*.

[2] Schiavone's etchings were very frequently based on Rosso's compositions.

---

appear in his own and in Matham's engravings), and thus the whole of the manner and subject-matter of the new Dutch mannerism and of the artists around Rudolph II, were based largely on Caraglio's engravings. Even the well-known chiaroscuro woodcut by Goltzius of *Hercules and Cacus* was based on Rosso, whose *Mars and Venus* was also the prototype for a whole series of erotic representations of the *Gods* by Goltzius: the cycles with famous war-heroes and the scenes after Ovid.

ninesque style of the forties and fifties was the Florentine painter Francesco Salviati. Though he cannot be dissociated from developments in Rome, his style showed an explicitly Florentine-mannerist intensity which enabled him —more decisively than Tintoretto—not only to preserve those propensities of later mannerism which were already contained within the earlier mannerist style, but to develop them further, so that the Florentine basis of the second, international mannerism seemed safeguarded during the transitional period of the mid-century and beyond. Typically Florentine were Salviati's wildly-jagged forms, the fascinating *élan* of his figures, the super-abundance of accessories (cf. frescoes of the life of *Furius Camillus*, Palazzo Vecchio, Florence). His manner of composition also was Florentine, deriving from Pontormo and Rosso. There were tendencies towards this type of composition about the same time (though in a highly personal way) in Michelangelo's late, 'mannerist', work, e.g. *The Conversion of St. Paul* (1542/45). Characteristic are Salviati's diagonally composed vistas of landscapes, and figures pressing obliquely into and out of the picture space (cf. frescoes of the *History of David*, Palazzo Sacchetti, Rome). From Rosso's frescoes in S. Maria della Pace and from Rosso's and del Vaga's *Loves of the Gods*, Salviati evolved his compositions of two intertwining nudes, still designed on a very shallow plane, and the complicated pose of the single nude figure (conceived partly in two, partly in three, dimensions). These he fashioned into a grandiose, rounded scheme. The symbolism and allegory of the figures are underlined by their artificial isolation: e.g. frescoes of the *Seasons* (Plate 6b) and of the *Creation* in S. Maria del Popolo, Rome,[1] and the various versions of the *Fall of Man*. This style had reached a stage of development not hitherto attained by the Northern artists, in spite of some attempts by Mabuse. It all looks like a foreshadowing of the ideals which the later mannerists were to strive after; it is not surprising if again and again it led to imitations, particularly in the circle around Goltzius. Salviati also extended

[1] These brilliant works, of such fundamental significance for Northern mannerism, as well as those in the Palazzo Sacchetti, are not mentioned by Voss, and so, apparently do not exist for art-history, notwithstanding the former include works by Raphael, Sebastiano del Piombo and Bernini in the Chigi Chapel—not after all a very obscure place. In this they thus share the fate of various other works to which I shall have occasion to refer in the course of this study.

and intensified the erotic element after the manner of Rosso and Parmigianino: the *raffinement* of a falling drapery or of an insecure piece of jewellery
(*Carità*, Uffizi) anticipated, word for word, the erotic-lascivious ideal of later
international mannerism. Another important link between the old and the
new style was the Tuscan artist Marco Pino. Historically, he grew out of
Beccafumi, as Salviati did from Rosso's and del Vaga's Florentine-Roman
style. Marco, in fact, worked under del Vaga. His typically Sienese, loose
manner of composition (*History of Alexander* on the ceiling of the Sala
Paolina, Castello S. Angelo) and especially his extravagant figures decisively
influenced the Netherlandish style at the turn of the century, chiefly no doubt
through his engravings. Marco Pino's types already looked almost like those
of Bellange, and he was in fact to influence Bellange as did no other Italian
artist.

### III

Thus Tuscan artists preserved the mannerist style during the transitional
period between the beginning and end of the century, and assured its organic
and intensive continuation. Everything points to Tuscany as the cradle of the
earlier Italian mannerism which in turn was to become the guide and prototype for later Netherlandish mannerism. All the elements, out of which
shortly after the mid-century the various patterns of the style were to
crystallize, can be deduced directly from the Florentine-Tuscan development.
It is wholly characteristic that the main Florentine artery was accessible to,
and invigorated by, related influences. This is of special interest here, from
the fact that these influences derived from precisely those artists who were
to be of great importance also for the North. In the 1540s the early compositions of Tintoretto appeared as the strongest parallels to Salviati and the
Florentine development.[1] His restless, excited arrangements go far beyond

[1] Tintoretto was at his most mannerist in his early paintings (and again in such late works
as his battle scenes) inspired probably by the wave of Florentine influence caused by
Salviati's and Vasari's sojourn in Venice (1539-41). It is typical of the still unclarified relationship of Tintoretto—an artist standing at the water-shed between baroque and
mannerism—to this Florentine style, that his important mannerist painting of the *Miracle
of the Loaves* (New York) has variously been accepted both as an early and as a late work
(the latter is probably correct).

the flat, decorative manner favoured by the North-Italian Parmigianinesque artists. As a Venetian, Tintoretto was closer to the baroque than any of his Florentine mannerist contemporaries. It is easy, therefore, to understand that, agreeably to the laws of cultural development, a mannerist artist in Venice should be able to draw the line of final demarcation between the two styles— for what else is the highly-exaggerated new mannerism, balancing so precariously on a razor's edge? To suppose, on the strength of Federico Zuccaro's work done in Florence in the mid-sixties and second half of the seventies, that it was he who introduced into Florence the Tintoretto-like type of composition, would be mistaken, notwithstanding that he was influenced from Venice. In fact, Federico's whole style of composition, including his early, smoother manner, and that of his brother Taddeo who, though the more important, died young, rested at every point on the compositional and formal idioms of Salviati and Marco Pino.[1] It was these two Tuscan artists who handed on the strong anti-realistic tendency inherent in earlier mannerism to the Zuccari, both of whom, through their paintings and the numerous engravings after them, were of such great importance for the North (Federico was himself in the Netherlands in 1574). And it was precisely by reason of this stylistic pre-conditioning that Tintoretto's method of composition, parallel to that of Florence, was able to influence Federico, and he in turn to transmit that influence, as an organic whole, to the Florentines. It is remarkable, moreover, that Federico, who frequently, indeed usually, gave the Tintorettesque type of composition a twist in the direction of Venetian classicism, should in Florence—no doubt under the influence of the *genius loci*—have been incomparably more mannerist than he was anywhere else, and have helped to advance existing tendencies yet farther. But the extent

[1] How closely Taddeo Zuccaro, and through him Federico also, was dependent on del Vaga (who furthermore was the source of inspiration for Salviati and Marco Pino) can be deduced not merely from the similarity of styles of the two artists, but also from the fact (not hitherto realized) that it was the young Taddeo who painted after drawings by Perino del Vaga, the frescoes of *Amor and Psyche* and *Perseus and Andromeda* in the Castello S. Angelo. Because of their mythological-narrative subject-matter and exaggerated style, they proved to be of great interest to Northern artists. Connecting stylistically the frescoes in the Castello S. Angelo and the cycles at Caprarola were the almost completely unknown works by Taddeo in the Palazzo Ducale at Castiglione del Lago on Lake Trasimeno.

to which the Florentines themselves (even in Rome) pressed the development of these tendencies, is manifest in many of the early works of Santi di Tito based on Salviati (e.g. the fresco *The Mocking of Nebuchadnezzar*, 1561/64, Vatican, Museo Etrusco). And here too there is much that anticipated Northern developments.

Only Florence with its mannerist tradition descending directly from Quattrocento gothic could have been the source of those mostly small pictures, usually of erotic, mythological and allegorical subject, which as its most charming and indeed most characteristic creation, one immediately associates with the general notion of 'Northern mannerism around 1600'. Thus the 'Studiolo', designed for the Grand Duke Francesco in 1570 in the Palazzo Vecchio, where hung close together and one above the other about thirty such pictures, crowded with graceful, engaging, artfully mannered nude figures, has already all the feeling, elegance and preciosity of style which marked the taste fashionable at the close of the century, particularly at the court of the Emperor Rudolph II at Prague. Specially characteristic 'pre-Rudolphian' pictures in Francesco's Studiolo are Allori's *Pearl Fishing*, (Plate I), Macchietti's *Baths of Pozzuoli*, Maso da San Friano's *Diamond Mines* and Poppi's *Continence of Scipio*, etc. From Borghini we learn what sort of pictures were chosen by other Florentine art patrons for their similar work-rooms: Botticelli, Piero di Cosimo, Pontormo, Stradanus, Netherlanders—no art-historical description could better illustrate the origin and development of the entire spiritual atmosphere of Florentine mannerism than this list of names.

Of the artists whom Borghini mentions it was above all Stradanus who, through half a century, constituted the connecting link between North and South, of which the importance can hardly be over-estimated. Born in Bruges, a pupil of Aertsen, he went first to Venice and Rome, then in 1554 settled in Florence where for a long time he worked as Vasari's assistant on the decoration of the Palazzo Vecchio. His extraordinarily fertile output was made known in the North chiefly through prints by Dutch and Flemish engravers. These consisted of historical, allegorical and mythological subjects, as well as numerous series and single scenes of religious, partly rather dogmatic, character, in which the artistic tendencies of the Italian and Flemish Counter-Reformation intermingle. No Florentine artist working in the sixties united within himself, as did Stradanus, predisposing qualities so apt for the formation, co-ordination and transmission of the gradually con-solidating apparatus of later mannerism. Attuned from the start to the Nor-thern undertones in Florentine art through his Netherlandish background, he found the point of contact with Salviati whose style was to be of the greatest consequence for European painting. (Traces of Salviati are discernible also in Vasari's works, though he used them in a more popularizing and relatively moderate form.) Stradanus combined the manner of Aertsen and Tintoretto with that of Salviati—under whom he had worked in Rome—thus contriving a synthesis of styles by which he in turn re-influenced the North. In the same way his references to Dürer, Baldung Grien, Bosch as well as to Rosso and Pontormo, fit in quite organically with Florentine art. The dependence of Stradanus' tapestries with hunting scenes (Plate 6a) on Salviati's drawings for his own tapestries of *The Seasons*, (Plate 6b) now becomes clear, as Steinbart observed. It was Salviati who, in the middle of the century in Florence, was the first to make resolute attempts at a new mannerist style. However, and contrary to Steinbart's view, Stradanus did not limit himself to a mere coarsening of Salviati's style; his compositions became more intricately mannerist compared with the tradition which he inherited on the one hand from Salviati and on the other from van Orley's Hunt-tapestries, as Steinbart correctly discerned. Now that we know the motivation of the development, we can mention also the impact made on Stradanus' series of Hunt-tapestries two years later[1] by Federico Zuccaro's large design for a theatre decora-tion of a Hunting scene (Plate 2b) (Uffizi)[2] executed for the Medici wedding of 1565, with its echoes of Tintoretto and Salviati. Steinbart sees Stradanus' influence solely as a designer of tapestry decorations and thereby fails not only to appreciate the historical importance of this artist for the whole of Europe, but, ignoring also the vogue his paintings and drawings

[1] The numerous hunting scenes by Stradanus were amongst the most widely distributed representations of the whole sixteenth century. Echoes from them, mainly through en-gravings, can be discovered, often in highly varied form, in Italian as well as in Nether-landish art, in the latter even as relief sculpture.

[2] On this very sheet Lindeman could have found the prototype of all the figures of *Actaeon* by Dutch artists and those working at the court of Rudolph II, namely the big figure of the Hunter on the right, with its typically Florentine and mannerist character-istics.

had in Italy, for his adopted country. On this, Borghini offers ample evidence. In any case, Stradanus designed no tapestries during the last thirty years of his life. The very painting of the *Mine* (Plate 4) which Stradanus executed for the Studiolo in 1570 anticipated, as perhaps no other work of the period, the compositional devices and even the figures of artists such as Bloemaert, Wtewael and Heintz. Tintoretto's grand manner was here reduced to smaller figures and a smaller format, and instead of the clearly articulated groupings of a crowd which were still to be found during the 1540s in Venetian-Parmigianinesque paintings, one is now on the whole confronted with sharply outlined, knobbly single figures or isolated groups bending vigorously forward or backward. Roughly the same relation existed between Stradanus's painting of the *Mine* and the landscape paintings of Pauwels Franck (Berlin), a slightly later artist, who lived in Venice, where he was called Paolo Fiammingo, and who also was of the greatest importance for Northern mannerism. Although his compositions have a marked similarity to those of Fiammingo, Stradanus seems a degree more linear, more accentuated and more pronouncedly mannerist, in key with the international trend. This resulted from his general reliance on earlier Florentine mannerist prototypes, especially on those which derived most directly from Pontormo's *The Martyrdom of St Maurice and the Ten Thousand*.

To any who may protest that the *Mine*, a typically 'Netherlandish' painting, was after all the work of a Netherlandish artist, I would answer that one has only to try to differentiate among the several hands employed in the decorations at the Palazzo Vecchio to realize how astonishingly close this particular picture stands to work of the Florentine Jacopo Zucchi; indeed, when they worked together in the same rooms not only are their styles virtually indistinguishable but also, as it seems to me, when they collaborated on the same frescoes. Zucchi, Florentine born, probably exerted the greatest influence on the Dutch mannerists who came to Italy, the more so as, with the exception of short interruptions around 1571 to 1587, he worked for a long time in Rome where many of his chief works are to be found. His speciality was pictures with mythological subjects of small format and with small figures, which have been mentioned before; but whereas hitherto painters had produced these only occasionally as a sideline, Zucchi raised them to the dignity of a distinctive class. Zucchi's pictures of the seventies and eighties show the Northern mannerist style of 1600 almost fully developed. Not only the court-art of the painters around Rudolph II, but the works of Wtewael, which are of special interest to us here, are difficult to conceive without Zucchi's paintings and drawings. Most marked in them are precisely those elements that are to be found also in Wtewael's small-figure paintings and in those of other Netherlandish artists: the same recumbent, crouching or seated small nudes (river-gods, nymphs, the members of Diana's court, and other such mythological figures) complete with elaborate jewellery, cunningly arranged to accent the erotic intent, with shells and corals, impossibly heaped still-lifes of flowers and animals, and rocky landscapes, in formations suggestive of antique landscapes of the *Odyssey* and of the earlier mannerism of Henri met de Bles. In these Zucchi landscapes, besides the light effects, one observes occasionally the tonality of Dutch artists such as Cornelis van Haarlem[1] and, above all, the composition and technique of Wtewael and Bloemaert.[2] However, not only was Zucchi the chief creator and popularizer of these small-figure paintings, he was also the originator of mythological frescoes in the Palazzo di Firenze and in the Palazzo Ruspoli in Rome (Plate 9a). The latter, among the grandest of cycles of later mannerism, had received only passing mention in the earlier sources, until Saxl recently drew attention to it and published it for the first time.[3] The ceiling of the Great Gallery shows the Gods of Classical Antiquity with their entourage, either as protagonists of the central situation of the myth connected with each of them, or as individual figures. This exuberant throng of figures must have produced an incredible impact on Netherlanders; it is as if the figures of a Wtewael or Cornelis van Haarlem had foregathered here before emigrating to the North. Without these frescoes, which represent the climax of Zucchi's *œuvre*, the whole formal, mythological repertoire of the

---

[1] It is almost inevitable, in the course of such a concentrated treatment of the mannerist style, that more should be said of composition and style than of colour. However, even a fuller consideration of colour and technique would hardly have affected my thesis with regard to the relationship of Tuscany and Venice.

[2] Moreover, Hans von Aachen's *Three Graces* (Braunschweig) correspond, as far as style and types are concerned, with Zucchi's painting (Coll. Czernin, Vienna) of the same subject. Also Verhaecht's *Golden Age* (engraved by J. Collaert) was obviously conceived under the inspiration of Zucchi's composition of the same theme.

[3] Fritz Saxl, *Antike Götter in der Spätrenaissance*, Munich, 1927.

engravings of Goltzius and his circle is unthinkable. The types of earlier
Italian mannerism, especially Rosso's single figures of the gods, which
Netherlandish artists often imitated directly,[1] re-appear on the Ruspoli-
ceiling in numerous extravagant variations. All the hitherto isolated poses of
earlier mannerism and of Rosso in particular, which Salviati had tried to
co-ordinate with but doubtful success, were here so easily and smoothly
interwoven and joined together that the Netherlanders must have discovered
in this masterly ensemble of figures, groupings, and properties an inexhaust-
ible treasure-house.[2] Zucchi's religious representations too were in line with
the aspirations of the Netherlandish artists, not only the large paintings in
S. Spirito in Sassia, but even more so the much more original, more pronouncedly 'Northern' frescoes in S. Silvestro al Quirinale of the *Adoration
of the Magi* and the *Circumcision* (Plate 5) from around 1575 (not mentioned
by Voss, and in the guide-books hitherto attributed to Raffaellino da Reggio).
These two frescoes show such close resemblances to Dürer as to amount
almost to copies; and some of the figures seem to be derived from Lucas van
Leyden, whilst the compositions are based on Stradanus. In no other work
of Florentine art is the Northern aspect so apparent as in these frescoes: the
archaizing types, the sharp-edged, angular and at the same time, imaginatively
arching forms. Here also was the same combination of styles that was to be
decisive for Flemish religious painting,[3] of which the works of Martin de
Vos and Joos van Winghe are the chief representatives.[4] Thus these Zucchi

[1] For example the engraving by Matham after Goltzius's *Summer* derives not only from
Zucchi, but also from Rosso's *Bacchus* (engraved by Caraglio).
[2] Only one example need be quoted: the whole *Bacchus*-group in the Palazzo Ruspoli,
also the figure of *Bacchus* in the Palazzo di Firenze recurs, slightly varied, in the Goltzius-
Matham engraving of *Ceres, Venus and Bacchus*.
[3] How much these works point to the direction which the development in the North
was to take, becomes evident from a comparison of the *Adoration* in S. Silvestro and the
*Adoration* by Brueghel (National Gallery, London), already strongly related stylistically.
[4] Thus there is an engraving by J. Sadeler after Martin de Vos derived from Dürer's
*Circumcision* (part of the series of the *Life of the Virgin*), which by that reason alone is
related to Zucchi's fresco. The engraving by G. Sadeler after Hans von Aachen comes
close to both the composition of de Vos and to Zucchi's fresco, from which the figures
of Mary and Joseph have been borrowed literally. The two engravings by Goltzius from
the series of the *Great Masters* (1594), deriving from Dürer and Lucas van Leyden, also
in their turn resemble Martin de Vos and the two Zucchi frescoes.

frescoes are eloquent of the important rôle which Florentine art played in forming an international style: through their direct derivation from Northern art, their connection with Pontormo, and thus indirectly with early Netherlandish mannerism; and finally, as parallels to the oncoming Northern mannerism, which depended equally on Dürer and Lucas van Leyden.

The development of Florentine painting—as seen in Stradanus and Zucchi—was strongly influenced also by Federico Zuccaro who, in the second half of the seventies, completed the frescoes of the *Last Judgment* (Plate 2a) in the cupola of the Duomo at Florence which Vasari had begun. These decorations were to be of the greatest importance for the whole of Northern art and, if only because of their size, should not have been overlooked by students of Netherlandish painting. Zuccaro did not merely associate himself with the trends then prevailing in Florence, he intensified them and—particularly in the lower part of the cupola—even surpassed all previous compositional solutions. As in the works of Wtewael and Bloemaert, the foreground looks like a stage-prop against which the Damned form a solid yet elastic trellis-work of contorted, tangled bodies. Sharper and more unreal than in Stradanus, and harshly accentuated by the light in the background, the rhythmic restlessness is emphasized through contrasted chiaroscuro effects. In the distance small, brightly-lit figures of the Resurrected, which correspond almost literally to those by Bloemaert and Cornelis, round off a composition which resembles closely what Northern artists were trying to do. (It is significant also that in both foreground and middle distance there are groups which seem to derive from motifs to be found in works of Signorelli.)

Thus the strongly uniform style of later Northern mannerism was, in all essentials, already formed in Florence between 1560 and 1570. Arbitrarily ornamental patterns controlled elongated, slender figures frozen in the most improbable poses. The figurative and spatial components no longer served to render a sense of reality, but, when respected at all, were exaggerated to such a point that they turned into its opposite. The groups of figures, played off one against the other, either surging forward or pressing back; the complicated, intertwining figures diagonally composed, the sudden gleaming vistas, were not meant to achieve a sense of unified space or the illusion of depth, but to create one co-ordinated two-dimensional ornamental net-work. This scheme—one might almost say formula—which had, as we know, already made a tentative appearance in the earlier mannerism as one of several possible methods of picture-composition, now appeared fully developed, as the last irrational extravagance of mannerism before the advent of the baroque.

As a notable example of the style we have been describing and as a sample of what Florentine art looked like at the end of the century in consequence of these developments, I would cite the religious and amatory mythological paintings and drawings of Andrea Boscoli, who both directly and indirectly was strongly associated with Northern mannerism. Through the impact of previous Florentine art Boscoli was led to a most elaborate ornamentalism, which was, to some extent, no doubt, a contemporary parallel to Northern art but, on the other hand, which was subsequently to influence Callot and especially the latest phase of Northern mannerism. This relationship to the North may be explained by his adherence to what was specifically 'Düreresque' in Pontormo, whose Certosa frescoes he transformed into a fantastic, veritably supra-Northern style, particularly apparent in his almost visionary drawings, comparable only with the wildest, most eccentric sketches of Bloemaert and more especially Wtewael. The flickering effects of chiaroscuro in these drawings, and his paintings, heightening to the utmost the unreality of the whole composition, indicate that Boscoli was influenced also by Barocci, who occupied an intermediary position between mannerism and baroque.[1] Barocci of course also made an impact on the artists of the Netherlands who, like Boscoli, interpreted his style in mannerist terms. The sharply segregated light and dark parts made their effect not in depth—a feature of Florentine mannerism before Barocci (see, for example, Rosso's *Christ carrying the Cross* in Arezzo)—but now under Barocci's influence, became interacting elements in a flatly composed rhythm.

Like Boscoli and Netherlandish mannerism, the Sienese artists Ventura Salimbeni and Francesco Vanni also interpreted Barocci's style in purely mannerist terms, possibly in an even more incisive manner. Their compositions show an extension and exaggeration of Beccafumi's principles and, in their abrupt contrasting of groups of figures and light effects, are akin to

---

[1] As an example of how Netherlandish artists interpreted Barocci's style, I would point to the transformation which his *Rest on the Flight to Egypt* received at the hand of Goltzius (engraving of 1589) and of Cornelis (engraving by de Gheyn the Elder, 1589).

Zuccaro's decorations in the cupola of the Duomo, and to the characteristic manner of Bloemaert and Wtewael. For centuries Sienese art had developed parallel to that of Florence, except that in Siena the Northern, spiritual, mediaeval element was always more important than in Florence.[1] Thus, in the period which is under consideration, the themes of Sienese art were mainly religious, in contrast to the secular and mythological subject-matter of Florentine painting. This Sienese devotional art found an echo also in the North, especially in Flanders, partly through engravings, often of markedly dogmatic content and character.[2]

IV

Now that the main outlines of Tuscan art during the second half of the sixteenth century have been discussed, it should not be difficult to define more precisely the historical positions of Frederick Sustris and Pieter Candid. Sustris, like Stradanus and Zucchi, took part in Europe's 'High School of Art': i.e. in the decoration of the Palazzo Vecchio under Vasari.[3] Consequently, as Steinbart has rightly stated, Sustris's style was decisively formed in what has been called the 'Vasari manner', as disseminated by Vasari himself and by artists such as Salviati, Gherardi, Stradanus, etc. Especially in his beginnings is Sustris to be regarded as a Florentine artist; even the Venetian element in his work, which he may have inherited from his father, or possibly acquired in Venice itself, underwent a Florentine transformation at his hand.

[1] Cf. F. Antal, *Gedanken zur Entwicklung der Trecento- und Quattrocento-malerei in Siena und Florenz. op. cit.*

[2] In this connection one can also point to the strongly 'Northern' etchings of religious subjects by Raffaele Schiaminossi from Borgo San Sepolchro, who attained a synthesis of Lucas van Leyden with Barocci and Salimbeni.

[3] Besides Stradanus and Zucchi, the chief participant in these frescoes was Battista Naldini, the pupil of Pontormo. There are a number of daring pen and wash drawings by him of religious and mythological subjects and of nudes, derived from Pontormo and Salviati, which also constitute forerunners of Dutch mannerism. The drawing of the *Judgment of Paris* which Kauffmann gave to Rosso, and which serves him to show the influence of Fontainebleau on Dutch mannerism, is in fact also by Naldini, and thus further demonstrates, if this were necessary, the Florentine impact on the North.

For example, in his tapestry of the *League between Florence and Fiesole*, dating from Sustris's Florentine days[1] the central figure is not Michelangelesque, as Steinbart summarily states, but resembles closely one of Salviati's favourite figures—it occurs in one of the *Camillus* frescoes, in the *Martyrdom of St. Lawrence* in the chapel of the Cancelleria, and in the *Beheading of St. John the Baptist* (S. Giovanni Decollato) (Plate 8b). Similarly, the standard-bearer in the same tapestry is a variation of similar figures in Salviati's *Beheading of St. John* (Cancelleria), and del Vaga's *Triumph of Scipio* (Palazzo Doria, Genoa). Precisely by virtue of his Florentine derivations, Sustris was already creating in the 1570s in the North (in Augsburg and on the Trausnitz) documents of later mannerism, works of exquisite grace and elegance, the like of which are unthinkable at that time in Haarlem or Utrecht, to say nothing of Fontainebleau, whither no Florentine had brought comparable works, so full of the *élan* and intensity characteristic of the source. In Bavaria, also, whither Sustris went sometime prior to 1569, the parallels and links, relating his style to that of the Florentine mannerists remained predominant, and there, quite independently, he created from Florentine-Netherlandish premises the international, so-called 'Goltzius' style, which was shortly to come to maturity throughout all Europe. Like Zucchi, Sustris extended and developed the stimuli received from Florentine art; that is why the frescoes on the Trausnitz—in front of which one might imagine one was in the Palazzo Vecchio—seem so akin to Zucchi's murals in the Palazzo Ruspoli. In the drawing of *The Triumph of Galatea* (Plate 8a), for the projected grotto-like hall for the east wing of the Munich Residenz, Sustris transformed in typically mannerist style Raphael's famous composition. The same had been done by Zucchi (fresco with the *Element of Water* in the Palazzo di Firenze) and by Spranger (engraved by Matham); and both Sustris and Spranger, in consequence of their Florentine training, produced what amounts to variations on Zucchi's fresco. It is characteristic that the types of the decorative female figures for this same grotto-hall should be reminiscent both of Vasari's (in the Palazzina Altoviti) and of Goltzius's *Judith* (engraved by Matham); and that Sustris's fresco, with the figures of *Athene* and *Ariadne*, in the Munich grotto-

[1] The *Parnassus* (Coll. Loeser, Florence), usually wrongly attributed to Stradanus, is probably also by the young Sustris, dating from his little studied Florentine period.

courtyard of 1587, should already suggest Bloemaert's early style of the 1590s.[1] The grotesques in Augsburg, Munich and on the Trausnitz spring from the Gherardi-tradition and resemble therefore most closely the contemporary ones by Poccetti. The figure of *Winter* (Trausnitz) is almost a replica of that by Salviati (S. Maria del Popolo); a general parallel to Salviati's compositions and figures emerges very clearly in Sustris's work. In the drawing of the *Triumph of Bacchus* (Louvre) there are many echoes of the figures in Taddeo Zuccaro's fresco *The Dance of Diana and her Court* in the Villa di Papa Giulio. Also the persistence of the fundamental earlier Florentine mannerism as the foundation of this development is evident in the Bavarian works of Sustris: for example, the figures of the gods in the Trausnitz decorations derived in general style, as well as in specific detail, from like figures—similarly placed in front of painted niches—in the engraving by Caraglio after Rosso. The influence of the sculpture of Cellini and Giovanni da Bologna on the designs for the statues of Evangelists and Apostles for the church of St. Michael is well known; but there are reminiscences too of Bandinelli (figure of St. Peter in the Duomo at Florence) and of Vincenzo de' Rossi in that both here and for the Trausnitz gods Marcantonio's engravings provide the basis. Sustris's *Christ carrying the Cross* (engraved by Custos) harks back to an older composition (used also by Veronese), which is related most closely to Pontormo's Certosa fresco and thereby both directly and indirectly also to Dürer. But Sustris exaggerated Pontormo's Florentine style into something much harder and sharper.

Pieter Candid, the other Netherlandish artist working at the Bavarian court, also studied in Florence and likewise took part, under Vasari, in the decoration of the cupola of the Duomo.[2] He also affected the international type of small Zucchi-like pictures as, for instance, in his *Venus and Aeneas* (Berlin). His tapestry designs of the *Seasons* (Plate 7b), done for the Bavarian

---

[1] Even if the composition is partly based on an engraving by Giorgio Ghisi after Primaticcio.

[2] A picture depicting the *Miracle of St. Nicholas of Bari* (formerly Coll. Contini, Rome) and eight charming lunettes with mythological scenes (formerly Coll. Acton, Florence) may turn out to be by Candid during his Florentine period. However, one could only be sure about such attributions after a thorough comparison with the certain, early works of the artist.

court, are specially noticed by Steinbart for the fact that they no longer represent abstract allegories but scenes from the life of the people. In this respect, they also have their mannerist ancestry in Florentine court art, for instance the *Labours of the Months* by Bacchiacca, designed as easly as 1552/53 for the Florentine court. Candid's series, half a century later, is reminiscent also of the *Seasons* by Stradanus engraved by Ph.Galle.[1] Steinbart suggests it was the influence of Barocci that later caused Candid to turn to the baroque; but this is only partly true and, as we have seen in the case of Boscoli and the Sienese artists, could very well have led in exactly the opposite direction. Another factor to be considered is the Venetian influence which increased throughout Europe around 1600 as the European trend towards baroque gained ground; formerly Venetian elements, even where directly conveyed, had been transmuted into a Florentine-Zuccaresque manner. The baroque tendency in Candid's panel paintings was, moreover, often based on baroque elements already inherent in the classical style of Andrea del Sarto and Correggio. Basserman-Jordan had already remarked that the altar-piece in the chapel of the Castle at Schleissheim is very close to del Sarto's *Madonna del Sacco* (an observation overlooked by Steinbart), and the *Madonna* at Oldenburg equally shows definite reminiscences of del Sarto. (Incidentally, Sustris's *Madonna with the Goldfinch*, engraved by S. Müller, comprises motifs from two *Holy Families* by del Sarto). Thus even in turning towards the baroque, Candid did not wholly deny his Florentine schooling and, by referring back to del Sarto, he demonstrated at this stage an historical development somewhat parallel to the phase in Florentine art represented by Jacopo da Empoli and Cigoli.

V

We now know more or less what were the formative conditions in which

[1] Antonio Viviani, the son-in-law of Sustris, who was called to Munich from Florence together with Candid and worked there with him (1585/92) was also clearly dependent on Stradanus and Federico's cupola in the Duomo, as can be seen in his *Allegory of the Old Dispensation* (1588/89) (Michaelskirche, Munich). Viviani is most probably identical with Barocci's pupil of the same name, who transformed the style of his teacher into mannerism.

later mannerism developed in Tuscany, and which of its representatives were to be of most significance for the North. The style appeared at its purest in Florence itself and from there directly influenced the many Northern artists who, on their way to Rome, stopped in Florence for longer or shorter periods. Nevertheless, as at the time of the earlier mannerism, Rome was the meeting place of all the leading, mainly Tuscan, artists, and at the same time the sounding-board and amplifier for the pitch to which all European art was attuned. Because of the powerful and positive impact radiated from this focus towards the North, we must now briefly consider more significant manifestations of later mannerism in Rome.

During the 1560s and 1570s Cardinal Alessandro Farnese, the most important patron of the arts of his time, commissioned decorations for his castle at Caprarola, near Rome. These were carried out by the brothers Zuccari, Giovanni de' Vecchi who came from the Tuscan borderland, Raffaellino da Reggio the pupil of Lelio Orsi, Antonio Tempesta,[1] Florentine pupil of Stradanus, Bartholomeus Spranger, and several others. The significance of these frescoes for the development and diffusion of the international style around 1600, may be likened to the rôle played by the frescoes of Simone Martini and his assistants, in the Palace of the Popes at Avignon, in the promulgation of that earlier 'international' style of the fourteenth century— though that from Caprarola was perhaps not quite as strong by reason of the less central position of the place.[2] Among the mythological frescoes by

[1] Netherlandish art owes such a tremendous debt to the rich treasure-house of Tempesta's inventions—chiefly transmitted through engravings—that an examination of his total influence would mean long-term, specialized research. I will mention a single example, because it is connected with the comparatively lesser known sources of religious painting in the Netherlands: Tempesta's spirited Florentine fresco of the *Massacre of the Innocents* in S. Stefano Rotondo (1580) corresponds in many details with Martin de Vos's composition, engraved by J. Sadeler. Probably the Flemish artist knew an engraving or a copy of the fresco, but theoretically, an influence of Martin de Vos on the pupil of Stradanus would also have been a possibility. (Both representations are of course transformations of Marcantonio's engraving after Raphael. The drawing by Swart van Groningen in the British Museum constituted an interesting intermediary stage.)

[2] Although, in contrast to those at Avignon, the frescoes at Caprarola have survived, they have not yet managed to rouse the interest of art historians. Apart from the successful attempts by Voss to clarify approximately the share of Taddeo Zuccaro and that of Raffaellino da Reggio nobody has yet taken the trouble to separate the various hands,

## The problem of Mannerism in the Netherlands

Federico Zuccaro at Caprarola, those in the Sala di Giove and Sala di Ercole tie up with the early, highly mannerist style of his brother Taddeo, best seen in his works at Bracciano, and, to a lesser degree, in the Villa di Papa Giulio. This early phase of Taddeo's was to be of great importance for Netherlandish artists. His lively narrative style of composition with its elongated figures and often Northern chiaroscuro, can be traced to the earlier Tuscan mannerism of del Vaga and Beccafumi. In Giovanni de' Vecchi who has not hitherto been evaluated the Tuscan roots of the mannerism of the Zuccari are more clearly exposed than in the work even of these artists themselves. De' Vecchi came from Borgo San Sepolchro and so was closely linked with Gherardi, the perfect collaborator of Vasari, who in many ways forms a parallel to Salviati. De' Vecchi's style has the typically Florentine, angular yet rhythmic forms, and clear connections with Quattrocento-gothic. His *Christ preaching in the Wilderness*, in the Sala della Solitudine, is like an interpretation, for Netherlandish eyes, of those figures by Rosso that look as if they had been conceived in imitation of wood-sculpture; and he approximates Zucchi's flow of line as closely as possible to the arbitrarily linear manner of the Northern style (compare, for instance, Zucchi's ceiling fresco of the *Seven Deadly Sins* in the Chapel of Pius V in the Vatican with de' Vecchi's fresco of the same subject in the Sala degli Angeli). *Gabriel appearing to the two*

---

let alone undertake a comprehensive examination of a work which is of fundamental importance for European art. Because any contribution to this highly important monument is useful, I shall put down some results, however provisional and incomplete, of a survey which was undertaken together with Roberto Longhi:

Sala dei Fatti Farnesi and Sala degli Concili: Taddeo Zuccaro (the latter mainly workshop and in parts by Federico). Chapel: collaboration of Taddeo and Federico Zuccaro. Sala dei Lanefici: F. Zuccaro (probably based on designs by Taddeo). Sala di Ercole, Sala di Giove: F. Zuccaro (probably based on designs by Taddeo). Sala della Primavera: F. Zuccaro (but not executed by him). The same applies probably to the Sala dell'Autunno and the Sala dell'Inverno. Sala dell'Aurora: F. Zuccaro (early style) and de' Vecchi. Sala dei Sogni, Sala degli Angeli, Sala della Solitudine, Sala dei Giudizi: d' Vecchi. The room with the landscapes: Raffaellino da Reggio. Sala delle Carte Geografiche: Raffaellino da Reggio, de' Vecchi, Tempesta. Decorations of the staircase: Tempesta. The little palace in the garden: completely decorated by Raffaellino.

One can predict that a thorough examination of the whole complex would elicit important art-historical surprises, especially with regard to the connection between Caprarola and the Netherlandish artists.

*Shepherds* and *Gideon and the Angel* (Plate 9b) (both in the Sala degli Angeli), constructed according to the Northern-Zuccaresque manner and with their appropriation of motifs and types from Rosso's *Labours of Hercules* already contained the fully-fledged type of composition of the later Northern mannerists,[1] and in the Sala dei Sogni (*Jacob's Dream*) parts of the landscape are completely in the Goltzius-Bloemaert manner. Whilst de' Vecchi comes very close to the Netherlandish artists in his way of composing, Raffaellino da Reggio has some incredibly Northern-looking types that might have come straight out of Bloemaert or Cornelis. These occur not only in his *Tobias and the Angel* (Borghese Gallery, Rome), but even more so in his decorations in the Garden House, the 'Palazzina' at Caprarola, particularly in the Loggia of the staircase. His astonishing *Ecce Homo* fresco (Plate 10) in the Oratory of S. Lucia del Gonfalone—this almost completely unknown 'Sistine Chapel of the Counter-Reformation' which was decorated by the Zuccari and their circle and, characteristically, also by Marco Pino da Siena—must have been of the greatest consequence for Flemish religious painting. In its own way, de' Vecchi's fresco reveals as strikingly as the works of the Tuscan Zucchi (with whom Raffaellino had collaborated in S. Silvestro) how much there was to be learnt by the North from Italian art of this time. The enthusiasm of the normally prosaic Baglione for Raffaellino and his significance for Italian art, is the Southern counterpart of van Mander's eulogy, quoted by Delbanco. Raffaellino's style, a synthesis of his Italian origins (Tuscany, Parmigianino, and the Zuccari), had an immense influence, not only inside Italy, but for that very reason, on the North also, now at last ripe for it.

Spranger, the decisive transmitter of the Tusco-Roman style (mainly via Goltzius) to the North, is of course to be explained first and foremost from the fact that he belonged in this Roman-Tuscan milieu which we have described; and which is further documented by his collaboration at Caprarola and by his relationship with Cardinal Farnese. The composition and figures in his chief work, the *Martyrdom of St. John* (S. Giovanni in Porta Latina) (Plate 11), show close affinities with Raffaellino's *Ecce Homo*, already mentioned, and with the frescoes of the story of *Hercules* by Federico Zuccaro at

[1] Amongst the engravings by J. Sadeler of Old Testament scenes after Martin de Vos, there are some which would more or less fit into this room; there is also much which corresponds with Cornelis van Haarlem.

Caprarola. The poses of many of Spranger's figures derive from Perino del Vaga (*Loves of the Gods*), from Bandinelli, and from Salviati (frescoes in S. Maria del Popolo). The two protagonists in his *Hercules and Deianira* (Vienna) are mere variants of the same subject in Caraglio's engraving of del Vaga's *Loves of the Gods*, and the figure lying on the ground in Spranger's picture comes from Taddeo Zuccaro's *Conversion of St. Paul*. Venus and Maia in the Vienna paintings *Venus and Adonis* and *Maia and Vulcan*, are both heightened versions of Bandinelli's *Cleopatra* (engraving by Agostino Veneziano). The famous, much copied *Wedding of Psyche* by Spranger and Goltzius is a transformation of Raphael's composition in the Farnesina, carried out with figures from Rosso and Zucchi. How closely Spranger's general style stands to that of Zucchi has been sufficiently emphasized.[1] His drawings, in technique and types, are often indistinguishable from those of Zucchi or Maso da San Friano. As with Zucchi, there are constant, strong reminiscences of earlier Northern and Italian artists. Among his types one finds, apart from Rosso and Giulio Romano, echoes also of Lucas van Leyden. He interpreted Dürer's *Adam and Eve* not simply from a solid sculptural point of view, as artists of the Mabuse-Bandinelli era had done, but, by elegantly accentuating its linear qualities, he created a work which recalls the late-gothic era (engraving by Goltzius, 1585). Spranger's interest in Quattrocento gothic is vividly illustrated by the fact that he made a copy of Fra Angelico's *Last Judgment*, as altarpiece for the tomb of Paul V, the grandest Pope of the Counter-Reformation.

I would draw attention here to another Netherlandish artist, the engraver Cornelis Cort, whose work in Rome between the years 1567 to 1570 and 1572 to 1578, though of great importance for the unification of the various elements in later mannerism, has not been considered hitherto in this stylistic connection, but only in relation to his Venetian activities. He made engravings after works by the Zuccari brothers, Marco Pino da Siena and Salviati, precisely the artists of greatest significance for the crystallizing out of the new style in Rome shortly after the middle of the century. Even in his engravings after artists of Michelangelesque tendency, Cort translated their sculptural,

[1] The shaggy sea-god in Spranger's *Glaucus and Scylla* derives from Zucchi (Palazzo Farnese) who took his type from Rosso (for example the engraving by Boyvin of the *Neptune* figure).

academic style into a Northern-Zuccaresque idiom. His work illustrates the artistic milieu in which Netherlandish artists lived and worked in Rome.[1] Cort also engraved works by Spranger, and had, like him, a considerable share in transmitting the Romano-Tuscan style to the North.[2]

[1] As an example of a direct line linking this art-historical focal point to the Northern artists, compare the engraving of 1567 by Cort after Taddeo Zuccaro's *Adoration of the Shepherds* (Plate 3b) (the original must have been executed earlier, as Taddeo died in 1565) with the *Adoration* by Hans von Aachen, designed for the Gesù between 1577-84, but known only from the engraving by G. Sadeler, and repeatedly copied in the North. This is a variant of Cort's engraving, but formulated more in the style of the later mannerism, and further enriched by half-length figures in the style of Marco Pino. Taddeo's version, by the way, goes back to an old Florentine scheme of composition which Bronzino developed in his *Adoration* (Budapest).

[2] El Greco, who was in Rome for some time at the beginning of the seventies, also had close ties with the artistic milieu just outlined. Like Spranger, El Greco had links with Giulio Clovio (Taddeo's collaborator at Caprarola) who had recommended him to the Cardinal Farnese, later patron of El Greco. Like Clovio, whose portrait he painted, Greco seems to have lived actually in the Palazzo Farnese. About this period of Greco's career there are several interesting remarks in Willumsen's *La Jeunesse du Peintre Greco*—a confused and amateurish book, but perhaps stimulating for that reason. Even if many of the attributions of Willumsen, a Danish painter, are not to be taken very seriously, it is not without art-historical interest, for example, that the half-length figure of *St. Dominic*, which Willumsen publishes as the earliest painting by Greco and which in reality is nothing more than a copy of an engraving by J. Sadeler after Spranger, does indeed show a stylistic relationship with Greco's *Magdalen* at Valladolid, thus providing a characteristic example of how, within the later mannerism, the limiting confines of these two artists could be blurred. The connection between Greco's works of his early Spanish period and the Roman, Tuscan, and Zuccaresque circles, his adoption and exaggeration of the late, roughly mannerist, Michelangelo (a fact which Willumsen also hints at), is obvious from many instances. Michelangelo's *Pietà* (Duomo, Florence) was the direct prototype for Greco's *Lamentation* (Coll. Huntington, New York), although far more exaggerated in the mannerist sense. The same work by Michelangelo was also the basis of *The Seat of Mercy* (Prado), but with the additional influence of Taddeo Zuccaro's *Entombment* (Borghese Gallery)—in itself a transformation of Michelangelo's group—from which the pose of the dead Christ and the assisting angels have been taken and exaggerated; furthermore, Greco made use also of Dürer's woodcut of the *Holy Trinity* in this composition. Spranger's *Pietà* (engraved by Goltzius, 1587) and Marco Pino's painting of the same subject (Villa Albani) also belong stylistically within this orbit.

   A further characteristic synthesis of Michelangelo and Dürer is Greco's *Derobing of Christ* (Munich): the crowded, ascending right half of Michelangelo's *Crucifixion of St.*

## VI

From this summary sketch of the complex development of mannerist painting it Italy, likely points of contact for Wtewael and Bloemaert clearly

---

*Peter* had conditioned the whole of Greco's composition and is particularly in evidence.
Especially the group of women, cut off by the lower margin of the picture, as well as
individual poses and gestures, were derived directly from Michelangelo, whilst the figure
of the bending executioner comes from the *Crucifixion* of Dürer's smaller woodcut
*Passion*. The inner connection of these two sources is further underlined by the fact that
Dürer, in the same woodcut, had already introduced the figure of the *Virgin*, cut off by
a hill, a motif which Pontormo elaborated in grandiose fashion in his Certosa fresco of
*Christ carrying the Cross*, and which Michelangelo had also used in the late frescoes of the
Cappella Paolina—taking it directly or indirectly from Pontormo. To have felt Dürer's
style as being close to that of Michelangelo's is a characteristic of earlier as well as of
later mannerism: examples are the sheet of studies by Pontormo (Uffizi 6702 F verso)
for the Certosa frescoes, which includes copies after both Michelangelo and Dürer, and
the famous engraving of the *Pietà* by Goltzius which represents a harmonious synthesis
of Dürer and Michelangelo.

The stylistic relationship between Greco's *Derobing of Christ* and the very 'Northern'
*Ecce Homo* of Raffaellino da Reggio is obvious. Greco's *St. Maurice* is a direct transformation and extreme intensification, in the style of the later mannerism, of Pontormo's
*Martyrdom of St Maurice and the Ten Thousand* (Pitti) which Greco might well have seen
in Florence. His *Resurrection* in S. Domingo (Toledo), is based not only on Titian's painting in Urbino, as is usually stated, but also—and specially in the way in which the composition appears as if torn asunder—on Michelangelo's *Conversion of St. Paul*, and either
directly or indirectly on Federico Zuccaro's *Resurrection* (engraved by Cort) and on the
same subject by Marco Pino (Gal. Borghese). (In spite of its supremacy, Michelangelo's
late mannerist style was nevertheless closely allied to the common style of his time,
pointing decisively to the future.) Greco's *Adoration of the Shepherds* in S. Domingo
(Toledo) derived—apart from the widely circulating engraving by Cort after Taddeo—
even more directly from another print by Cort after Marco Pino. In the same way as
Titian had inspired Greco's *Resurrection*, so Bassano (*Pentecost*) had stood sponsor for the
*Adoration;* but the Venetian elements, particularly those coming from Bassano, were
transformed in the same manner as Wtewael and Bloemaert were to adopt later. The
Roman parallel to El Greco is probably most clearly evident in some of the works of
Giovanni de' Vecchi. Furthermore, it is characteristic of the connection between Greco
and the sphere of Tusco-Roman mannerism, that Latanzio Bonastri from Siena, a pupil
of Beccafumi, should have become his pupil in Rome. From all these details emerges the
new awareness that Greco, beyond his contacts with Venice, had close links with the
whole artistic circle we have outlined; of his partial dependence on Italian prototypes;
and his appurtenance to the style of the later mannerism.

suggest themselves. Evidently earlier students of Italian mannerism, looking for Italian parallels to the particular artist under consideration, have been content, through unfamiliarity with the general setting, to fish out more or less at random the first names they happened on. Thus, for instance, Lindeman, estimating the impact on Wtewael of his Italian tour, equates all that is Italian in him with Bassano. And it is true, of course, that Wtewael did make use of Bassano, but by no means in his entirety as one integral artistic source. Thus firstly he limited himself specifically to Bassano's mannerist, Parmigianinesque style; then secondly he carried this style further in the direction of that uniform, generally approved method of composition and vocabulary which was characteristic of later mannerism, whose real home, as we have seen, was Tuscany. Similarly in Venice, it was overwhelmingly from the most expressly mannerist, and therefore basically Tuscan, works that Wtewael drew inspiration.[1] A good example, showing how the Northern artists—even in Venice—discovered, or instinctively copied, Northern-mannerist elements as something intrinsically akin, is Wtewael's *Adoration of the Shepherds* (Vienna) which, with its hard, sculptural, chiaroscuro effects and in many of its details and nuances, is much closer to the painting of the same subject by Nadalino da Murano (1554/55, Sacristy of S. Sebastiano, Venice)[2] than to the *Adoration* by Bassano in Vienna which Lindeman re-

[1] As an example of how Venetian mannerism and the ingrained mannerism of Tuscany intermingle—and how even Titianesque motifs were being translated into mannerist terms by Northern artists—I would cite the two pictures of *Diana and Actaeon* by Wtewael (Vienna and Coll. Prince Heinrich) which probably derived from Paolo Fiammingo (engraving by G. Sadeler)—who himself shows similarities to Zucchi—as well as directly from Florence, namely from Tempesta's engraving *Diana and Callisto*. The point of departure for most of these works was of course the relevant compositions of Titian.

[2] Nadalino's picture, like Wtewael's, with its nocturnal light (the source of light being the Christ Child) makes such a 'Northern' impression that Wickhoff was led at one time to attribute it to Martin de Vos. This shows that Greco could have taken this 'Northern' motif from Venice, or from the Loggie and not only from Correggio, as has been maintained. That Wtewael's connection with Nadalino's painting was not fortuitous, is evident from another related work: Hans Speckaert's *Adoration* (known from an engraving by G. Sadeler), an artist who worked in Florence and Rome and who was close to Hans von Aachen. The paintings of Nadalino and Wtewael both contain mannerist transformations of motifs from Raphael's *Sacrifice at Lystra*. There had of course been adaptations of this tapestry design by Raphael during the period of the earlier mannerism, for example

produces for comparison, and which no longer belongs to that artist's mannerist period. On the other hand, one of the most mannerist figures of Bassano's Parmigianinesque period, the *St. John the Baptist*, occurs more than once in Wtewael's *Adorations*, its sharpness exaggerated. Yet these *Adorations* from Bassano's mannerist period tended to be very flatly composed. For the conscious, or unconscious, source of Wtewael's much more restless *Adoration*, it may be necessary to look to Verona where, as in Tuscany, a constant Northern bias prevailed and where the transformation of Venetian mannerism—for example in Farinati's and Zelotti's works[1]—was akin to that of the Northern artists. The mannerist elements that Wtewael drew from North Italian art had much the same effect on him as those from Tuscany itself (from Zucchi, for example), where Bassano's Parmigianinesque manner and, for the most part, Wtewael's style also originated. This merging of North Italian with the manifold influences of Tuscan, and particularly Sienese, art (Zucchi, Boscoli, Vanni, Salimbeni), seems to have been achieved by Wtewael without difficulty; it can best be seen, for instance, in the *Holy Family* at Gotha. The Florentine influence is most clearly apparent in Wtewael's characteristic mythological paintings with small figures; and the impact of Zucchi is obvious in the *Judgment of Paris* (Budapest) (Plate 12b). This influence emanated not only from Zucchi's small paintings, but also from the large frescoes in the Palazzo Ruspoli, many of whose deities are easily to be recognized in the works of Wtewael, even where they have undergone much alteration. Closely connected with Zucchi's influence is that of Spranger; both made a direct impact on Wtewael, and also indirect

---

[1] Also the mannerist style of the young Veronese is only to be understood from the traditions prevailing in his native Verona. It was this early style which impressed itself most forcibly on the Northern artists.

---

an *Adoration* by Polidoro da Caravaggio which, typically enough, was engraved by the school of Goltzius. Furthermore, most renderings of the theme of the Adoration after 1567—from Wtewael to Rottenhammer (painting in Vienna)—contained echoes of the engraving by Cort after Taddeo Zuccaro which for its part contained mannerist translations of motifs from the *Sacrifice at Lystra* and of other works by Raphael. Another *Adoration* which is dependent on Cort's engraving is that by G. B. Pozzi in S. Maria Maggiore, whose illustration in Voss's book had probably induced Lindeman to suggest a parallel between this artist and Wtewael.

through their influence on one another. Wtewael's early painting of *The Flood* (Germanisches Nationalmuseum, Nuremberg) (Plate 12a) is typical of later Dutch mannerism. Viewed solely from within the native tradition it is probably to be judged as based on Heemskerck; but Italian derivations abound, though—as always with Wtewael—nothing is copied exactly and one finds a combination of many stimuli. This large composition with its nudes, sharply contrasted in chiaroscuro, recalls generally Zuccaro's paintings in the cupola of the Duomo at Florence; but there are traces also of Tintoretto and Salimbeni. It is a highly mannerist interpretation of the subject and may be said to represent the culmination of a series starting with Michelangelo's *Flood* and proceeding, via Bronzino's *Passage through the Red Sea* (Cappella di Eleonora, Palazzo Vecchio), to Allori's *Pearl Fishing* (Studiolo, Palazzo Vecchio) (Plate 1). Among the single figures one finds transpositions from Michelangelo (Sistine Ceiling figures, particularly the Slaves and Jonah), and exaggeration of some of his characteristic motifs (particularly that of gripping the hair), already noticeable in Bronzino (*Martyrdom of St. Lawrence*, S. Lorenzo, Florence) and in the frescoes in the University of Bologna by Pelegrino Tibaldi, a Bolognese artist in close contact with Tuscan art.[1] The children in Wtewael's picture are reminiscent of Tibaldi. The figure

[1] A comparison between Wtewael's drawing of *The Death of Adonis* (Albertina) and Tibaldi's *Polyphemus* fresco (University, Bologna) strikingly reveals the connection between the two artists—one can almost certainly speak here of an influence by Tibaldi on Wtewael—although both artists referred to Michelangelo's *Venus* for their figures. Because of this relationship, a work such as Wtewael's drawing *The Baptism of Christ* (Albertina) is reminiscent (in spite of its Sienese-Florentine elements, cf. Jacopino da Conte's identical subject in S. Giovanni Decollato) of the Italian followers of Tibaldi, such as G. B. Crespi (*Baptism*, 1601, Frankfurt). The importance of Tibaldi for the North is also evident from a comparison of the figure of *Icarus* by Cornelis van Haarlem (engraved by Goltzius) and Tibaldi's *Slave*, mentioned before. This figure is so near to Tibaldi—nearer even than to Cort's engraving after Titian's *Prometheus*—that one is inclined to think the Dutchman must have known copies of the Tibaldi figure.

I should like here to enlarge the *œuvre* of this artist, so important for the Netherlandish mannerism, by two outstanding works. In the Sala Paolina in the Castello S. Angelo, not only the *Archangel Michael*, which has been attributed to him, is from his hand but also the large-scale, bronze-coloured scenes from the History of Alexander the Great, some of whose figures seem to point forward to Cornelis. The scenes are fitted into ornamental frames with painted cartouches, scroll-work and garlands, above *bassamenti*

of the youth in the right margin of the painting seems to come from the boy raising his arm in Pontormo's fresco-lunette at Poggio a Caiano, but here translated into Zucchi's Palazzo Ruspoli-manner; and even the nudes in the foreground seem like echoes from Pontormo.[1] From the way in which these Italian impressions have imprinted themselves, it is possible to arrive at a natural dating for *The Flood* and the Dresden *Parnassus*[2]; *The Flood*, with its intense exploitation of Italian mannerist stimuli, cannot have been painted, as Lindeman maintains, before Wtewael's Italian journey, but soon afterwards, i.e. after 1592, whilst the disputed date of the *Parnassus* picture in which Wtewael employs his final economical method of composition, is probably to be read as 1596, not 1590, as Lindeman considers plausible.[3] Wtewael's drawings especially reveal his surprisingly parallel relationship with Boscoli. They often show the same doughy, exaggerated style as the Florentine, and the same figures looking like a tangle of twisted roots, driven together as if by a fierce gust of wind (e.g. *The Adoration of the Magi*, and the *Annunciation* in the Louvre). Salimbeni's *Andromeda*-drawing (Munich) offers an equally marked parallel to the style of the Dutchman. Other sheets, such as the *Judgment of Paris* (Braunschweig), the *Banquet of the Gods* (Dresden) and the *Diana and Calisto* (Amsterdam)—whose attribution to Wtewael, however, I find somewhat doubtful—answer almost exactly to the nudes in Taddeo Zuccaro's *Amor and Psyche* fresco.

Thus one can discover in Wtewael's work extraordinarily many and direct connections with Italy which the artist established during his stay here. Al-

---

[1] The child in the right corner of the *Susanna and the Elders* (Gouda) derives from Andrea del Sarto.

[2] With regard to the *Apollo* in this painting, Fischel has pointed to a drawing by Raphael in the British Museum. (Cat. *Raphael and his Circle*, No. 29.)

[3] As far as the chronology of the two paintings is concerned, I find myself in agreement with Stechow (review of Lindeman's book in *Zeitschrift für bildende Kunst*, 1929/30).

---

in whose rectangles sea-gods have been fitted. The single figures of Antique Gods between columns and almost all other decorative figures in the room are by Sermoneta. In the library next to the Sala Paolina, Tibaldi also executed the magnificent friezes with sea-gods, which hark back to Rosso's Parmigianinesque style, whilst at the same time pointing forward to Zucchi and Spranger. In the attribution of these two cycles, which up to now have been given partly to del Vaga, partly to Marco Pino, Roberto Longhi agrees with me.

though he used them in manifold combinations, they appear quite natural to us once we realize their origin. Bloemaert, on the other hand, never went to Italy. He represents a wholly Northern parallel to the Italian development. One reason for the similarity in stylistic development of the Utrecht artist and the Italians, was Bloemaert's transformations of Barocci (whom Bloemaert knew of course from engravings, and whose *St. Francis* he copied in a drawing now in the Albertina). With its unreal spatial effects achieved by means of flickering chiaroscuro, with its marked ornamentalization, it is most nearly related to comparable Tuscan transformations of Barocci at the hands of Salimbeni, Vanni and Boscoli (whom Bloemaert possibly knew also through engravings). If Delbanco, confronting such a surprisingly Boscoli-like picture as *Judith before the People* (Vienna) is reminded of Tintoretto, this is merely another example of the sort of vague guesswork which should now be abandoned for a more precise clarification of artistic relationships. How right one is, when considering European mannerism, to stress the persistence and constant renewal of the earlier in the later style, is shown by the fact that even in Bloemaert, who was acquainted only indirectly with contemporary Italian art, and can hardly have seen any example of earlier Italian mannerism, this style should yet have been, as it were, involuntarily revived: for instance in his earliest painting, the *Niobides* (of 1591, Copenhagen) (Plate 14), the style of Rosso's *Moses and the Daughters of Jethro* (Uffizi) spontaneously and unconsciously breaks through.

## VII

Hitherto we have been concerned mainly to discover the important points of contact in Italy for the Netherlandish style throughout the century. I should like now to show how painting in the Netherlands itself developed, from its own indigenous traditions, in the direction of Tuscan art. Here I enter upon territory explored by Otto Benesch who, in his full and important introduction to the catalogue of the Netherlandish drawings of the fifteenth and sixteenth centuries in the Albertina[1] was the first to attempt to clarify both the overall development of Netherlandish painting of the sixteenth century, as

[1] *Beschreibender Katalog der Handzeichnungen in der Graphischen Sammlung Albertina. Band II: Die Zeichnungen der Niederländischen Schulen des XV. und XVI. Jahrhunderts. Bearbeitet von Otto Benesch*, Vienna, 1928.

well as its constant connections with Italy. In the first place, the critical publication, with illustrations, of all the Viennese drawings, is of great art-historical interest; I would wish, however, in this context, to confine myself to an examination of Benesch's account of the historical development which, in spite of a conciseness inevitable in an introduction, is scientifically the most considered, precise, and important study of the subject yet produced. I shall therefore examine Benesch's arguments closely and combine my remaining observations with some criticism of his study.

I hope I interpret Benesch's complicated analysis correctly, when I say that he distinguishes three phases in the development of Netherlandish painting of the sixteenth century: (1) the spiritualized, unsculptural style of Netherlandish mannerism about 1520 (for example, the group around Bles); (2) Romanism, the break-up of the Northern manner, with ever stronger links with Italy, and the gradual gestation of an international style. This protracted development extended, according to Benesch, from Mabuse to Floris, Stradanus and Blocklandt, and included Netherlandish artists working at the courts of Prague and Munich during the century; (3) the mannerist style at the end of the century, beginning, say, with Martin de Vos. National differences between Flemish and Dutch art now came to the fore, and in contrast to a preoccupation during the second period with purely formal problems, there was a return to a concern more with spiritual content. And there were signs also of a reorientation toward the early mannerist style.

The sharp differentiation and indiscriminate use of the terms 'mannerism' and 'Romanism', a terminology adopted by Benesch from previous students of the period, obscures, from the outset, the true, exceedingly complex picture of sixteenth century art in the Netherlands. 'Romanism' is a general historical concept (like 'Renaissance'), which in this art-historical context, should refer only to what is in the highest degree Italianate in Netherlandish sixteenth-century art: thus strictly, only to such classically orientated artists as Orley and Lombard. Used as a term analogous to, or more comprehensive than, mannerism,[1] or as the designation of some impossible, monstrous style

---

[1] What for instance is the meaning of: 'his contemporaries and followers provide stronger evidence even than Floris himself, that the new phase of Romanism is concomitant with a new inflowing wave of mannerism. After the middle of the century, Romanism bears a mannerist stamp'?

stretching from Mabuse by way of Floris to Spranger, it is meaningless. This misleading use of the term 'Romanism' rests on an inadequate appreciation and analysis of the relationship between Netherlandish art and Italy; basically on the fiction of one constant, 'Italianate art', valid for the entire sixteenth century in the Netherlands. But the situation in Italy itself was too complex, the inter-relationship throughout the period too persistent, and at the same time too variable, to warrant treatment as a single style, such as this so-called 'Romanism'. The accents and tensions lie quite otherwise; they are too varied for it to make sense to settle for one comprehensive style, to include, for example, Mabuse, Lucas van Leyden, Pieter Coecke and Heemskerck, of which 'Italianate' shall be the common denominator, with but occasional mannerist nuances. One should speak rather of the mannerism of Orley, and of Lombard, as being occasionally tinged with Roman classicism.

It is quite impossible to interpret the difference between mannerism and Romanism at the beginning of the century as if the latter were Italianate and progressive, the former un-Italian and reactionary. Contradicting his own thesis, Benesch at one point states that the earlier kind of mannerism of the group around Bles, 'belongs already to a new period'. The real foundation of this observation he is unable to give; it would clash at the very outset with his entire thesis. If this style still shows a flamboyant, late-gothic flavour, it is a mannerist reaction to the, sometimes stronger, sometimes weaker, Italianizing, Leonardesque style of David and Matsys—a reaction comparable to that which took place at the same time in Italy, and which was likewise accompanied by a revival of late-gothic trends (Pontormo-Botticelli, Rosso-Filippino).[1] This is forgotten by those who assert the conservative character of the Bles group, in contrast to the contemporary, in fact slightly earlier, 'Romanistic' style of Mabuse. As I said at the outset, this is the stage in which it is far more difficult to draw a clear distinction between pre-Renaissance, late-gothic and post-Renaissance mannerism in the North, than it is in Italy

---

[1] The dependence of Mabuse on Filippino and of the early Orley on Signorelli is well known (see Ch. Aschenheim, *Der italienische Einfluss in der vlämischen Malerei der Frührenaissance*, Strasbourg, 1910). One can recognize how Orley's *Job*-altarpiece (Brussels, 1521) grew out of Signorelli's fresco of *The Last Judgment* at Orvieto, and thus presents a kind of parallel to Michelangelo's *Brazen Serpent* (Sistine Chapel) and Rosso's *Moses and the Daughters of Jethro*.

where the cleavage made by classicism was much more clearly marked and effective. Benesch's concession that there may have been some kind of connection between the early Mabuse and mannerism, is meaningless, for it does not take fully into account the important large nudes, or the *Venus* in Rovigo (Plate 15)—this late-gothic, mannerist transformation of the 'Medici' Venus—which looks like a Botticelli in Northern terms, and in which the late-mannerist style of Spranger is foreshadowed to such an astonishing degree. The parallel, Mabuse-Rosso, as it is revealed in the comparison between the *Hercules and Omphale* (Coll. Cook) and the frescoes in S. Maria della Pace, is beyond question, and spotlights clearly Mabuse's historical position. With both Rosso and Mabuse the nude figure is developed as an end in itself, and becomes the protagonist of each picture; but in Mabuse there is of course still quite a lot of Renaissance influence, whereas Rosso had already turned away from the Renaissance.[1] Mabuse's 'Romanism' and the 'mannerism' of the other Antwerp artists were complementary roughly in the same manner as were the styles of Pontormo and Rosso, or as the sculptural and ornamental tendencies within the mannerist style. In Italy from 1525 to 1530 the emphasis on 'volume' increased steadily and was dominant throughout the forties; essentially the same development occurs, each after his fashion, in many Netherlandish artists, and is still discernible in Lucas van Leyden. His earlier, late-gothic type of mannerism illustrates well how it was that Bacchiacca, the Florentine artist, came to borrow from Lucas the Northerner. Lucas's later work veered in the direction of Mabuse-Rosso-Bandinelli.[2] To maintain Benesch's classification it would be necessary to call Rosso and Bandinelli 'Romanists', and to interpret the development in Florence and Rome from the end of the 1530s to the beginning of the 1540s as an evolution from mannerism to 'Romanism'.

During the period between the two very pronounced mannerist styles, one at the beginning, the other at the end, of the century, which Benesch would fill with a single 'Romanistic' style, there existed in fact in the Nether-

[1] Compare Mabuse's *Neptune and Amphitrite* (Berlin) and Bandinelli's *Adam and Eve* (Bargello) which both derive from Dürer's engraving of *Adam and Eve*.
[2] Apart from its origin in Mabuse, Lucas van Leyden's engraving *Lot* seems already to contain references to del Vaga's and Rosso's *Loves of the Gods* cycles, from which one print was also copied at this period by the Master of the Female Half-lengths.

lands several stylistic trends which eventually lead to the second mannerism. To list and accurately distinguish the several waves, parallel to Italian art, one must be mindful of their diversity; sometimes they are developments rather of the native tradition, sometimes more decidedly of trends deriving from Italy. This period of transition is most clearly marked by the following chronological sequence of names: Heemskerck, Floris, Blocklandt. Their works show on the one hand the transitional rôle of these artists—similar to that of Salviati in Italy—as well as their gradual advance towards the corresponding stage which the development in Italy had reached (with Blocklandt one can almost speak of a catching-up). This development, characterized in Italy by the works of Salviati and Marco Pino, was accomplished in the Netherlands under much more intricate and diverse conditions, owing to their totally different origins. In the North it proceeded by fits and starts and required the endeavours of several different artists, both contrasting and complementing, the later often taking over where the earlier left off.

During the forties and fifties in Heemskerck, and occasionally in Aertsen, one finds certain parallels to, as well as consequences of, the Parmigianinesque, Schiavone-Bassano style, which we have recognized as an important link in the continuity of mannerism around the middle of the century. However, owing to his Northern, unbalanced and violent character, Heemskerck's style appears much more vehement than anything that was being done in Italy at the same time.[1] Owing to his close connection with the native art of the North (Engelbrechtsen, Bles), Heemskerck attained in the Netherlands, during the middle of the century, the rôle of an exponent and preserver of mannerist intensity. In this he may be considered a Northern parallel to Salviati, with whom he probably worked in Rome in 1536 on the decorations for the *Entry of the Emperor Charles V*. One need only remark how incredibly involved and ornamental is, for example, his *Conversion of St. Paul* (engraved by Coornhert) in comparison with its obvious prototype: Salviati's fresco in the Cancelleria. As this mural was executed after Heemskerck's Roman days he can have known it only from a copy. Or again compare his imitations of

---

[1] Benesch calls the style of this artist 'Romanistic baroque'. What is that supposed to be? Such lighthearted and fanciful designations produce only unnecessary confusion.

Rosso, Bandinelli[1] and Giulio Romano in the series from the Old Testament (also engraved by Coornhert). And even the relatively tame *Crucifixion* of 1543 (Ghent) shows the same stage of development as the contemporary *Deposition* in the Prado by Jacopino da Conte, a Florentine associate of Salviati. Heemskerck made available to the Netherlandish artists a storehouse of forms and themes which he presented with tremendous gusto, in a wildly-ornamental and fantastic style, closely connected—as we have said—with that of the group around Bles, and at the same time reminiscent of the extreme mannerism then in vogue in Italy, particularly evident in Beccafumi, whose 'Northern' types are often remarkably like those of Heemskerck.[2] And in many of his compositions, particularly those with large foreground figures, as it were, plunging into the picture-space, there is an affinity with the pre-Bellange style of Marco Pino. Furthermore, engravings and drawings by Battista Franco are important links between North and South, and between the earlier and later mannerism. Contemporary with Heemskerck yet independent of him, these show the same types and the same quivering, ornamental, super-Bandinellesque way of rendering the Antique. This profusion of Heemskerck's Italian, more specifically Tuscan, mannerist characteristics and achievements, with their typical poses and style, their rich store of stooping and stumbling figures, provided, not surprisingly, many prototypes for later mannerism. The whole ragged, doughy style of Wtewael, Salimbeni, Boscoli, the Bellange-types, as well as many of Blocklandt's and Cornelis's figures, are foreshadowed here. Ph. Galle's engravings of the *Last Judgment*, with the dark wreath of figures silhouetted in the foreground and brightly lit scenes of small figures behind, point forward to Stradanus and Zuccaro's paintings in the cupola of the Duomo at Florence.

In contrast to Heemskerck's generally flat compositions, one finds in

[1] The direct connection of the almost super-Florentine Heemskerck with Florentine mannerism, underlines at the same time the relationship with Florentine Quattrocento Gothic. For example, Heemskerck's *Judith* (engraved by Coornhert) stems from Donatello's famous group.

[2] How naturally these wild 'Northern' Heemskerck-types evolve within the Italian tradition (Giulio, Bandinelli, Rosso) can be seen in Agostino Veneziano's engraving of Raphael's *School of Athens* which appeared as early as 1524.

Aertsen,[1] as a consequence of his development and heightening of elements present also in earlier Netherlandish mannerism (e.g. Scorel[2] and especially Pieter Coecke), definite tendencies parallel to the Tintorettesque-Florentine style of composition found in later mannerism (particularly noticeable in his representations of the Crucifixion) (Plate 13a). Precisely because he was so rooted in his Northern origins (from Jan van Amstel), Aertsen easily assimilated the rhythmic and ornamental characteristics of the Italian style of the 1540s which derived from Parmigianino, Schiavone and Bassano, whilst in his forceful, mannerist narrative power he sometimes approaches the remarkably 'Northern' manner of Polidoro da Caravaggio,[3] one of Raphael's pupils (e.g. *The Raising of Lazarus*, Coll. Lugt).[4]

During the 1550s a somewhat calmer climate prevailed both in Italy and in the Netherlands. The exaggeratedly mannerist, highly unbalanced northern *élan* of the Heemskerck-Aertsen circle was followed by a period of restraint, in which more deliberate attempts at imitating Italian art were the rule. Floris was the chief exponent of the new trend. At the beginning, in the *Judgment of Solomon* (Antwerp) (Plate 17a), for example, one discerns the influence of Raphael, of antique sculpture, of the relief-like compositions of Bandinelli[5] (*Martrydom of St. Lawrence*) and of the classicizing mannerist art

---

[1] The giant still-lifes by Aertsen and Beuckelaer, with one or two figures, are deceptively similar to those by the Cremonese painter Vincenzo Campi.

[2] By underlining the forward-thrusting movements that had already been latent in Michelangelo's *Flood*, Scorel reveals in his painting of the *Pentecost* (Utrecht) his adherence to what was later to become the manner of composition of the later mannerism.

[3] It seems an art-historical miracle that a pupil of Raphael was able to execute a painting so much in the style of Heemskerck and Aertsen as Polidoro's *Christ carrying the Cross* at Naples.

[4] Thus Stradanus could absorb from his first teacher not only a heightening of his own native tradition, emanating from Pieter Coecke, but also much that was inherently Italian, when he went to Venice and Florence.

[5] Lambert Lombard, the teacher of Floris, is the most obvious Netherlandish parallel to Bandinelli as well as to Battista Franco's Roman style. How closely he himself felt his relationship to be to this whole group, is shown by the fact that he copied Rosso's *Massacre of the Innocents* (Drawing, British Museum, catalogued as by 'Rosso'). He was also influenced by Giulio Romano whose style, significantly, he intermingled with that of Mantegna—Vincent Sellaer, also a member of the generation before Floris, is another artist who revealed Italian influences, although in a little more retrograde fashion. At the outset he imitated the Lombard followers of Leonardo, a trend common to many

of Giulio Romano.[1] He had probably known Salviati in Rome, but only on his return to the Netherlands did Floris's compositions, and even his types, begin to approach in vitality those of the Italian painter. A good example is the *Fall of Man* (Pitti), dating from 1560: it is highly reminiscent of Salviati, though as far as the individualization of the figures is concerned, he did not go beyond Rosso. Viewed from within the native Netherlandish development the painting seems of course to be based on Mabuse. The various cycles by Floris with single female figures (Nymphs, the Five Senses, etc., all engraved by Cock) must surely have been conceived partly under the influence of Salviati's *Seasons* (S. Maria del Popolo), although both in the handling of drapery and the landscape motifs they betray many more regional ties than would have been the case had they been painted by Salviati. The composition also is more two-dimensional, not based primarily on fore-shortened figures as with Salviati. Apart from Salviati's impact on Floris one detects also the closely related influence of Vasari. And how nearly the influence of the early Tintoretto, which was of special importance for Floris, can coincide with that of Salviati is demonstrated in his picture *Susanna* (Gal. Ferroni, Florence). Through Giulio Romano, Floris was linked with the sources of both Venetian and Fontainebleau mannerism, and in his evolution towards the style of Salviati was for a time actually helped by that of Fontainebleau, to the foundation of which both Giulio Romano and del Vaga had contributed, and where the development had likewise tended towards an admittedly very moderate version of Salviati's style.

[1] The heads remind one of those that occur in the frescoes by Jacopino del Conte in S. Giovanni Decollato.

---

Netherlandish artists, as Steinbart recognized in his essay on Sellaer (*Münchner Jahrbuch*, 1928). Later on Sellaer veered towards Florentine art, and—typically of the slower development of the North—echoes from del Sarto and Bacchiacca, also from the early Bugiardini (*Leda and the Swan*) can be discerned. Apart from the oft quoted borrowings from Raphael and Michelangelo, this whole generation leaned heavily upon Florentine mannerism. For example, Hemessen's *Holy Trinity* (Erlangen) repeats almost exactly the type of St. Anne in Bacchiacca's and Bronzino's *Holy Families*. Moreover, there is a copy by Hemessen after del Sarto's *Caritas* in the Scalzo Cloister (formerly Coll. Bozzi, Vienna) in which del Sarto's style was transformed into the more sculptural manner typical of the circle around Bronzino—a proof that Hemessen must himself have been in Florence.

Kauffmann justly observes that Floris had strong links with Fontainebleau —but this has never been questioned. However, before discussing Floris further, we must consider the development of the Fontainebleau style itself, after which it will become clear when, and in what manner, links with the Netherlands were possible, and when out of the question. The Netherlandish artists did of course seek active contact with Italy by all possible means during the very complex transitional period around the middle of the century; and it goes without saying that Fontainebleau, this neighbouring enclave of Italianate art, was made full use of. However, as Benesch agrees, the main influence was Italy itself; therefore not too much should be made of the connection with Fontainebleau. It is characteristic of Fontainebleau that the stylistic battles of the Italian homeland were mirrored there in much milder fashion, that hardly or only occasionally does one see it as reflecting the great artistic development in the South. Within the aristocratic and conservative framework of Fontainebleau, many mannerist elements lost their bite and force. Rosso's spirited Florentine style of the 1530s—mild compared to his earlier works, and made widely familiar chiefly through engraved drawings —was followed during the next decade by Primaticcio's slightly Raphael-esque-Antique manner, derived from Giulio Romano and late Rosso, and which can be called—in a general sense—Parmigianinesque,[1] though Prima-ticcio was never as pointedly mannerist as the Italian artists of the same period. During the 1550s, he again adopted a strong Raphaelesque, more classical style, and in this sense closer to del Vaga.[2] Very close to Primaticcio also was Niccolo dell'Abbate who, during the 1550s and 1560s revealed at times even more pronounced Parmigianinesque traits which, in his more extreme works, he transformed into a del Vaga-Salviati-like style. For an appreciation of the whole so-called First School of Fontainebleau, one must bear in mind how incomparably stronger and more expressly mannerist was the contemporary work of the Florentine Salviati, derived from the same elements: Rosso,

---

[1] In a somewhat modified form this also applied to Dumûtier.

[2] Whilst Rosso's impact was at its greatest during the sixteenth century, Primaticcio's influence, characteristically, was felt most strongly in the seventeenth century, and par-ticularly in the circle around Rubens.

Parmigianino, del Vaga, and Giulio Romano.[1] After a considerable lapse of time there followed the Second School of Fontainebleau (Dubreuil, Dubois, Fréminet). Here features of later mannerism began to appear towards the end of the century—though in much more moderate form and later than the second mannerism in the Netherlands and, moreover, under the influence of both Italy and Flanders. Dubois and many others were Flemish; Fréminet was a completely Italianized Frenchman.[2] Fontainebleau, which is supposed to have been the source of later Dutch mannerism, began as an offshoot of Italian art, and ended as one of the very last outposts to fall into line with the international style at the close of the century. It was from the outset a reservoir of classicism, and the French artists were far more classically orientated than the Italians who, although they may have set the tone, tended to be more conformist.[3]

Arising from this account of the development at Fontainebleau, Floris's connections with French court-art can be documented by a comparison of his *Feast of the Gods* of 1550 (Antwerp) with Primaticcio's *Concert* (Louvre) or the two versions of the *Caritas*, one by Floris (Coll. Semanoff, Leningrad), the other by dell'Abbate (Coll. Grab, Vienna)—examples could easily be multiplied—all of them deriving from Italian sources: the former from Giulio Romano, the latter from Salviati's *Caritas* (Uffizi). Floris's *Labours of Hercules*, engraved by Cock, also refer back to Italian sources, resting, as they do, on Rosso's series which were still being produced in Italy. Compared

[1] It is significant that Salviati, during his stay at Fontainebleau around 1555, did not transmit anything of Italian mannerism.

[2] Even Dimier admits the Second School's complete dependence on Italy.

[3] A direct and impressive confrontation of the French art of Fontainebleau and its contemporary Italian counterpart is presented by the decorations in the Palazzo Sacchetti in Rome, where around 1553/55 the French painters Ponce (who in Meudon and probably also at Fontainebleau worked together with Primaticcio) and the future court-painter Marc Duval created frescoes next to Salviati's extremely mannerist cycles. The appearance side by side of this colony of artists, originally completely dependent on Italian mannerism, and of Italian artists themselves, provides a contrast which makes art-history suddenly and dramatically come alive. The way in which French classical art differs from its Southern mannerist counterpart, how Ponce's frescoes, in spite of a certain similarity with Parmigianino, point straight to Poussin, becomes there completely convincing and makes the Palazzo Sacchetti the ideal spot to prove the truth of such stylistic theses, and is to be recommended to any adherent of the Fontainebleau hypothesis.

with Heemskerck and Coornhert, Floris transposed Rosso into a clearer, more linear, even if more ornamental, style.[1] A detailed comparison of these two cycles would provide an extremely instructive contribution to the history of Cinquecento art.[2] But Floris did not stand still. His painting of the *Sea-gods* of 1561 (Stockholm) (Plate 13b) which was of fundamental importance for later mannerism in the Netherlands, intensified the style of Giulio's *Birth of Venus* in all its motifs and details (shells, fishes), and showed the possibility of a Northern line, running by way of Spranger to Wtewael, parallel to the Florentine line Salviati-Vasari-Zucchi. And, of course, the Stockholm painting was a translation also of Raphael's *Galatea*, similar to that of Zucchi and of Spranger, who was himself influenced by both Floris and Zucchi. The rising Italian mannerist wave of the 1560s (still revealing here and there the Parmigianinesque Schiavone-Bassano style of twenty years earlier) gave a new impetus to Floris's own mannerist tendencies, and to the associated manner of composition derived from Florentine sources as well as from Tintoretto.[3] Floris's renewal of the ornamental forms of Heems-

---

[1] This cycle provides an interesting stylistic parallel to the sculptures of the same subject by Vincenzo de'Rossi (Palazzo Vecchio), which were executed a little later and which in their turn make use of Rosso and Bandinelli in similar fashion.

[2] Aldegrever's *Labours of Hercules* of 1555 contain many reminiscences of the cycle by Rosso and at the same time of Valvassore, Pollaiuolo, and Robetta, and of course also of Dürer. This involved relationship becomes even more complicated, but at the same time clearer, owing to the parallels between Dürer and Pollaiuolo, and between Rosso (*Hercules and Achelous*) and Dürer (*Samson*)—Jacob Blinck, the man who called Cornelis Floris to Königsberg in 1549, also engraved the whole series of Rosso's figures of deities.

[3] In the same decade, Breughel produced the grandiose drawing of *Summer* (1568) which demonstrated his adherence to a manner of composition that was derived from Florentine mannerist sources, as well as from Tintoretto, and which is akin to a drawing by Stradanus, mentioned before, which was based on Aertsen, Tintoretto and Salviati. Also of interest here is the relationship between Brueghel and Giulio Clovio, the greatest miniaturist of his time, the collaborator of Salviati and Taddeo Zuccaro, and an intimate friend of Cardinal Farnese, the patron of Greco. Students of Brueghel have overlooked Giulio Clovio's inventory of 1577 in which four works by Brueghel are mentioned. The subject-matter of three of these works can be looked up in the inventory, the fourth I shall quote here, as it is of significance for the connections between Italy and the Netherlands at that time: 'Un quadretto di miniatura la metà fatta per sua mano, l'altra di Mo. Pietro Brugole.' It is possible that Clovio simply coloured Brueghel's drawing—as he had done with Heemskerck's *Triumph of Charles V* series (British Museum)—but it is

kerck and Aertsen was dressed in that Parmigianinesque style which changed colour with every decade, at the same time becoming progressively sharper.[1] As with Heemskerck, one detects in Floris the influence of engravings by Battista Franco, Battista del Moro, and Giorgio Ghisi. The various tendencies of the transitional period in North and South, having originated in different traditions, now drew slowly closer together: the main forces in the Netherlands were directed towards the formation of a truly international style, which in Italy was already fairly well developed, but which can only really be called 'international' after it had reached full maturity at the end of the century. This is what we mean when we refer to the 'later' or 'second' mannerism. But where and what is Benesch's unified international style shortly after the mid-century, with Floris as its central figure? It can neither be identified nor understood through his formula of 'a synthesis of Raphael, Michelangelo, and Parmigianino'.[2]

[1] Benesch makes the following correct observation: 'The artistic tendencies which gained expression in the late-gothic exuberance of the 1520s, also pervaded the style of Parmigianino and his circle, which adopted it as something very much akin.' Why then has Benesch interpreted the mannerism of the 1520s as something specifically un-Italian? This repeated contradiction arises from his basically faulty viewpoint which does not consider Europe as a whole. The Italian sources remain obscure because Benesch considers Michelangelo as the chief influence on Lombard, Heemskerck and Floris. It is far too generalized and therefore misleading to quote Michelangelo as a direct source, because the question is not whether the Netherlandish artists borrowed anything from Michelangelo, but in what manner. Not Michelangelo himself, but the interpretation of his works at the hands of Rosso, Bandinelli, Giulio Romano, Franco, Salviati, Vasari, etc., was their source. (The same applies also in the case of Raphael. I have therefore usually indicated the Italian transformations of Michelangelo's and Raphael's works rather than the originals whence most of the motifs stemmed.) It is typical for example that Michelangelo's drawing of *Phaeton*, in which his connection with Florentine mannerism is particularly evident and which Salviati recreated as a painting, was so frequently copied by artists of the Netherlands: particularly by Floris, who exaggerated the inherent ornamental qualities of the drawing. ((Drawings in Berlin and in the Louvre); by Martin de Vos (engraving by Ph.Galle); by Hans von Aachen (painting in Castle Ambras), etc.)
[2] After completion of this manuscript there appeared the work by Dora Zuntz on *Frans*

equally possible to interpret the entry of the inventory literally, especially as it was compiled in Clovio's presence, and to see in this drawing a work of collaboration between the two artists. Bredley's suggestion that the two artists had been friends was probably based on this assumption (*The Life and Works of Giulio Clovio*, London, 1891).

The third artist of great significance for the transitional period is Anthonis Blocklandt, who is here put forward as the chief representative of an entire group. He it was who now united previous efforts in the Netherlands directed towards realization of later mannerism. His extraordinary mannerist intensity is to be explained by his development out of the late Floris, by his connection through Floris with the rising Parmigianinesque wave of the forties, and at the same time by his reference back to Heemskerck. His relationship to Salviati was only of a general nature: his reconstruction of Salviati's style came closer to the original than did Floris who interpreted him in a more 'classical' sense.[1] Blocklandt carried the native development so far that it was at last possible for the Netherlands to catch up with Italy, and for the later mannerism style to establish itself in the North. Blocklandt too had connections with Fontainebleau, though more with Rosso than with Primaticcio. Compare, for example, Blocklandt's mythological narratives (engraved by Galle) with the *Dance of the Dryads* (engraved by Boyvin). But Blocklandt showed a strong tendency towards the new mannerism (which at that time had not yet caught on at Fontainebleau) by translating the Rosso-Primaticcio manner into a style something like a cross between del Vaga and

---

[1] *The Youths in the Fiery Furnace* (Haarlem) by Pieter Pietersz, another artist of Blocklandt's generation, corresponds so closely in motif and overall style to the *Camillus*-frescoes, that it seems likely he knew them at least from copies.

---

*Floris* (Strasbourg, 1929). This contains an *œuvre-catalogue* which is useful, although not very well arranged and not always correct. For example, it seems incomprehensible why the *Feast of the Gods* in Antwerp, dated 1550, should not be by Floris and why date and signature are supposed to be false. The author provides neither an account of the development and position of the artist as viewed from the Netherlands' point of view, nor does she establish in any methodical way the connections with Italy. She talks mainly about borrowings from Michelangelo, Tintoretto and Giulio Romano, and mentions occasionally Salviati and del Vaga. The connections with Rosso and Bandinelli have not been recognized, and the relationship with Fontainebleau are denied. The frieze of the *Seagods* in the Castello S. Angelo, which Floris might have seen whilst it was being executed, is by Tibaldi (see note 1 page 79), not by Giulio Romano, as the author copied from her Anderson photograph. The decorations of the Palazzina Altoviti are not by del Vaga, as Zuntz repeats after Venturi, but by Vasari. Probably because the author has not examined the historical importance of Floris's development thoroughly, her choice of illustrations is rather confusing; the works of the highest quality are missing.

Taddeo Zuccaro (cf. *The Dance of Diana and her Court* in the Villa di Papa Giulio). The painting of *Venus and Amor* (Coll. Nostitz) demonstrates how much sharper was his style than that of either Floris or dell'Abbate, whose *Caritas*-compositions it resembles. And this painting shows also how Blocklandt took from Primaticcio what was most Parmigianinesque in him, and so manipulated it that it now resembled the early, still mannerist, Veronese (the *Venus* especially reminds one of Veronese). On the other hand, in this figure of *Venus*, the intensely mannerist mode of del Vaga and Parmigianino was to some extent reaffirmed, thus providing a link with earlier Italian mannerism. This link was achieved via the Genoese artist Cambiaso (himself much influenced by del Vaga), on whose famous *Venus and Amor* composition—widely publicized by replicas, copies and engravings—Blocklandt must have modelled his picture. His composition of the *Resurrection* (engraved by Ph. Galle) shows how closely Blocklandt had assimilated also the Tusco-Roman style, say of S. Lucia del Gonfalone. There we find the same heightened interpretation of Michelangelo's frescoes in the Cappella Paolina, the same affinity of style with Marco Pino, and the same direct relationship to Cort's engraving after Federico Zuccaro that we noted in the case of El Greco, who was in Rome at the same time as Blocklandt.[1] A similar parallel to El Greco may be observed in the works of Dirck Barendsz, a contemporary of Blocklandt who resided for many years in Venice. His *Christ expelling the Moneylenders from the Temple* (engraved by J. Sadeler) is 'Bassano turned into mannerism', thus recalling El Greco's early painting of the same subject (Coll. Cook). It is, moreover, of particular interest that Stradanus's composition of this theme had obviously also made an impact on Barendsz, so that here—as so often with Netherlandish artists—Venetian, Tuscan mannerist, and Northern elements mingle. Blocklandt's works show traces also of North-Italian mannerist artists: the early Veronese, Zelotti, Farinati, Cambiaso, and especially—as later with Wtewael—the engravings of Battista del Moro. Block-

---

[1] Blocklandt's figure of the Sentinel who has fallen backwards—this typical figure which also occurs in Greco's *Resurrection* and which derived from the Cappella Paolina and from Tintoretto's *Miracle of St. Mark*—was probably known to both artists from Taddeo Zuccaro's more elaborate rendering of the same figure in his *The Taking of Christ* (S. Maria della Consolazione). The recumbent figure on the left in Blocklandt's picture appears to be a reminiscence from Signorelli.

landt's stylistic connections, indirectly with Floris and North Italy, and directly with Parmigianino[1]—at a time when a decorative-ornamental Parmigianino-like style was prevalent in the Netherlands, which in its turn resembled the style prevailing in Italy during the 1540s—paved the way for, and indeed already included, elements of Spranger, Bloemaert and Wtewael; even Blocklandt's figures are occasionally quite Bloemaert-like.

To clarify this development from Floris to Blocklandt, one should compare the *Judgment of Solomon* by Floris (Plate 17a) with Blocklandt's *Joseph before Pharaoh* (Centraal Museum, Utrecht) (Plate 16a). The basis for both was Raphael's *Blinding of Elymas* and Bandinelli's *Martyrdom of St. Lawrence* (and for Floris in addition the *Judgment of Solomon* from the Loggie and Bandinelli's *Massacre of the Innocents*). But how different were their respective interpretations. Floris's composition stuck fairly close to its models, combining classical elements with a statuesqueness derived from Bandinelli—the latter quality being softened by the former. This can best be seen in the figure of the Executioner holding up the child; he borrowed this from the *Massacre of the Innocents* and changed into a stocky, heavily draped Roman. Floris was not yet rife at that stage of his development for a mannerist interpretation of this figure; it was Cornelis who, in his *Massacre of the Innocents* (Haarlem), first took over the figure from Bandinelli literally. The female costumes—à la Heemskerck—in Floris's painting are, however, no less fanciful than those in Bandinelli's picture. And where Floris's composition has a well-balanced appearance, Blocklandt's designs seem restless and unstable. He must have looked at the work of the early—still mannerist—Veronese, for example the curvilinear composition of *Christ disputing in the Temple* (Prado); whilst in the background there are already signs of some Zucchi and Dürer types. His handling of space—wide and sweeping, yet at the same time unrealistic—was anticipated in some of Heemskerck's works. Finally, Floris's *Judgment of Solomon* was the starting-point of a line of mannerism which ended in Blocklandt's excitingly composed second painting of the

[1] Benesch makes the rather generalized, but essentially correct remark: 'The chief element that left its stamp on the art of the Netherlands during the third quarter of the sixteenth century was Parmigianino's influence. It made itself felt even in works which stemmed from the Venetian orbit.'

*Lot* series (engraved by Galle)[1]—both pictures were based on Raphael's *Blinding of Ananias*. Blocklandt's *Temptation of Lot*, the fourth of the series (Plate 16b), is directly related to Floris's composition of the same subject (Plate 17b) (also engraved by Ph. Galle). Like the Italian mannerists, the Northern artists (Lucas van Leyden, Floris, Blocklandt, and Wtewael) put ever greater emphasis on the erotic element in this part of the *Lot* story—one reason why the theme was so popular in the North. Blocklandt's style[2] contained, basically, almost every ingredient of the fully matured later mannerism, except that it was still too powerful, too massive: the final touch was yet to come direct from Florence, from Florentine delicacy and its capacity for abstraction.

VIII

The overall stylistic development which followed Blocklandt, and the way in which later Florentine mannerism assumed such a predominant position in Europe and brought about a completely international style, has been outlined earlier. Benesch, owing to his mistaken reasoning, denies the internationalism of later mannerism—one of its most pronounced characteristics—and differentiates sharply between the style as it emerged at the end of the sixteenth century in Flanders and in Holland respectively. He speaks of a 'division in what had been internationally fused' (*Scheidung des international Verschmolzenen*), thereby turning undeniable differences of nuance into a difference in essence. He takes the view that the later Dutch mannerism is a foreign body, grafted on to an autonomous growth. This is as one-sided and erroneous as the opposite view, held by Kauffmann, which today is the more

---

[1] The third of the *Lot* series, as so often with Blocklandt, was based on Parmigianino, in this case the *Abraham and the Three Angels* (known only from an engraving). Similarly, Blocklandt's *Martyrdom of St. James* (Gouda) is dependent on the *Martyrdom of SS. Peter and Paul* (engraving by Caraglio) whilst the *Adoration of the Shepherds* (Göttingen) derives from Parmigianino's *Adoration* which again has survived only in an engraving.

[2] Blocklandt's compositions also proved a stimulus to German art. The *Lamentation* (Ludgeri Church, Münster) by Nicolas tom Ring, which Hölker illustrates as the chief work of this artist in his book *Die Malerfamilie tom Ring*, Münster, 1927) is nothing more than a copy of Goltzius's engraving after Blocklandt.

commonly accepted. Both views fail to take into account all the ramifications of the sixteenth century; nor do they recognize the natural, and therefore indivisible, intermingling of the indigenous and the adopted elements in later mannerism. Ultimately even Benesch cannot deny the international status of those artists who, according to him, worked in their different 'national' styles at the court of Rudolph II at Prague and in Munich. But owing to his artificial thesis he is obliged to tear them out of their chronological and stylistic context and to attach them arbitrarily to his 'Romanistic', international, formalistic epoch of the mid-century.[1] Contrary to his own thesis, Benesch states elsewhere that the style of the artists around Rudolph II was 'a new and original language of their own, developed independently, whereas the earlier generation had been influenced from abroad'. But it would really seem impossible to separate the style of the artists working in Prague from that of the Haarlem painters—one has only to think of the indissoluble relationship of Spranger and Goltzius (which Benesch of course also realized)—or to ignore the links between the Munich court-art and Florence. One cannot ascribe two distinct styles to the Netherlandish artists who worked at the European courts, and to those who stayed at home: to the former an international, to the latter a national, style. Furthermore, one cannot overlook the fact that the Netherlandish artists working in Italy identified themselves with the style of their adopted country in a way that would not have been remotely possible a few decades earlier. What we are confronted with at this time—and precisely at this time only—is (apart from slight individual differences) a truly European style which, at the end of the century, represented the climax of an overall, parallel, cosmopolitan development. Only when one recognizes the fundamental connection between later mannerism and Florence—which, as we have seen, far from hindering the

[1] In order to resolve this contradiction, Benesch presents a very contrived picture of the development. He speaks of an international style which reached its height shortly after the middle of the century, and he maintains that it originated from artists who were born around 1520 and worked chiefly abroad or at foreign courts. However, most of these artists belonged almost exclusively to the next generation of 1540 to 1570. According to Benesch, the Flemish artists returned from Italy at the end of the century, after a short apprenticeship, the evidence for this being what he calls their 'national' style. However, when he excludes those who did not return because of the Wars of Religion or because they remained at foreign courts, he invalidates his theory as to their return.

autonomous development of the North, spurred it on—and applies it correctly, does one hold the key to the manifold problems of the later mannerism.[1]

The harking back at the end of the sixteenth century to the earlier mannerism, the revival of native gothic elements, did not represent a local reaction on the part of the Northern painters, but was a manifestation typical of the later, international mannerism, typical not only for the North, but recognizable also in the Florentine-Roman form of mannerism, and thus a vital and all pervading phenomenon for this international style. Benesch does not recognize the close connection between the neo-gothic elements in the North, which of course he admits, and the decisively strong Italianism, which he vigorously denies.[2] Though he allows to his 'national' Dutch artists a return to the earlier form of mannerism, a concession which he has also to make in the case of the 'international' artists at the court of Rudolph II, yet he does not see that Spranger, for example—precisely because he was a 'late-gothic' artist—laid himself open to direct Italian influence; nor does he see that it is one and the same phenomenon that both the artists at the court of Rudolph and those in the Netherlands should base their art on their common native

[1] According to Benesch's theories, it would appear that the Venetian rather than Florentine influence was stronger with the artists at the court of Rudolph II and on Flemish artists generally. In fact the chronology of the stimuli was rather the other way round: first came the impact of North Italy, and only then did the influence of Florence make itself felt. To generalize one might rather exaggeratedly say that the artists created something North-Italian out of elements they imported from Italy, even if these came from Florence; whilst later they made something Florentine, even from elements which came from Venice.

[2] For example, the revival of 'gothic' at Augsburg, also in architecture and sculpture, at the end of the century, the very time when the connection with Italy was alive and artists who had studied in Florence were working at Augsburg (Hans Reichel; the fountains by Adriaen de Vries; Hubert Gerhard): and the same phenomena which were observed at the beginning of the century—absorption and transformation of gothic elements of the Quattrocento into mannerism—repeated themselves, *mutatis mutandis*, even more strikingly. (This natural, and art-historically fascinating, development would make a perfect rider to my Breu and Filippino article. Zeitschrift für Bildende Kunst, 1928.) It is relevant also in this connection to mention the Munich sculptor Hans Krumpper, who modelled figures of the Apostles from designs by his father-in-law Frederick Sustris, but at the same time also made gothic monstrances.

traditions. On the one hand, he speaks of 'conscious archaism' (*Historisierung*) as one of the hall-marks of mannerism yet, on the other hand, he maintains that the Netherlandish artists of that century, whom he wishes to invest with a special 'national' style, went 'beyond conscious archaism and found in the era of Lucas van Leyden a vital stimulus'. Once again Benesch seeks to demonstrate the principle of a Dutch national style by a slight shift in the nuances, thereby distorting the overall picture. The question whether the revival of the older style in the Netherlands during the sixteenth century is a matter of 'conscious archaism' or 'vital stimuli' is such a complicated one that it can only be answered from case to case, and then hardly ever un-equivocally.[1] But on no account may these conceptions be used as generaliza-tions or even as catch-words. The turning of the later towards the earlier mannerism, with its late-mediaeval elements, pinpoints a certain gothic gracefulness in both. Forming, as it did, the stylistic basis of the whole of the forthcoming European art, this gracefulness could have originated only in an Italian city such as Florence which had so many gothic and Northern elements in its artistic tradition. Here, the earlier mannerism appeared so graceful and delicate, because of its connections with the gothic style of the Quattrocento (in Venice no such phenomenon arose). For this reason the whole development cannot be correctly assessed without taking due account of Florence as the seminal centre.

I have tried to elucidate, by means of a mosaic of detail, the style of the later Dutch mannerists such as Goltzius, Cornelis, Wtewael, and Bloemaert. We have seen how the North-Italian, Parmigianinesque style was gradually displaced by the sharper Florentine one. This development can best be illus-trated in the work of Goltzius. Through his master Coornhert he still had connections with Heemskerck but, by slowly repressing the predominantly

---

[1] For instance the backward glance towards the earlier Netherlandish mannerism, to Rogier, to the Master of Flémalle and the adherence to the style of the Master of the Virgo inter Virgines. Lombard's interest in the Italian primitives, in Margaritone d'Arezzo, and at the same time in the mediaeval stained-glass windows of his own country, which he mentioned in the important letter to Vasari, was primarily of an historical and scienti-fic nature. See also Panofsky's essay on the 'historical' attitude towards gothic, *Städel Jahrbuch* VI, 25–72, 1930 [English ed. in Panofsky's *Meaning in the Visual Arts*, New York, 1955.]

Northern elements, he achieved in the end almost complete identification with Italy.[1] The fact that the young Goltzius found his point of contact with Heemskerck, Stradanus and Blocklandt, shows that his own development had prepared the Haarlem painter to receive and assimilate the fully matured style which Spranger transmitted to him around 1585. Even the traces of the older Tuscan mannerism, which Spranger had also passed on, are apparent in Goltzius's early period.[2] Through Coornhert and Heemskerck there were in his early period links with Rosso; the engraving of *Honor and Opulentia* (1582) (Plate 18b) is very close to one of the series of del Vaga's *Loves of the Gods*, and though still strongly like Blocklandt in manner, yet points already towards the elegant and elongated figures in his engraving of *Adam and Eve* after Spranger. The very Rosso-like *Jupiter and Europe* print, done after the entry of Spranger's style into Haarlem, is already completely in the new international style. Thereafter one meets in every field successful manifestations surprisingly similar to Italian work: mythological scenes close to Zucchi, and religious representations in the manner of Boscoli and Salimbeni (Goltzius-Matham *Christ and the Woman of Samaria* (Plate 18a)). The light palette of Dutch painting is often strongly reminiscent of that used by Tuscan artists: Bloemaert's almost rococo-like tones resemble those of the later Sienese artists; the flickering colours of Cornelis, those of Beccafumi[3]; whilst the often smooth, enamel-like and colourful Wtewael recalls the Florentine artists.

From the overall viewpoint of European artistic development, and in light of the earlier, and thus highly influential, parallel Italian movement, Flemish painting represents but one special nuance of the international style around 1600, and particularly of Netherlandish art. Flemish art, like its contemporary Dutch counterpart—and in spite of sometimes considerable stylistic differences—had, nevertheless, equally strong links with Italy. Stradanus, who

[1] How much Heemskerck's Northern, super-Florentine style became Italianized after half a century—by toning down the Northern elements and heightening the Italian—is shown by a comparison between his *Adoration of the Shepherds* (engraving by Coornhert) and the closely related print by Stock after Bloemaert of the same subject.
[2] A drawing at Berlin shows how exactly he copied Bacchiacca's painting of *Adam and Eve* (Coll. Johnson, Philadelphia, formerly Frizzoni, Milan).
[3] Observation of Stechow.

influenced the North as well as the South at one and the same time, bears witness to the close attachment of Flanders to Italy and in particular to Florence. Within limits, Benesch appreciates the importance of Stradanus[1] though not his rôle as the moving force in the Florentine development that, from the start, was so closely allied to the North. Stradanus fits as naturally into Florentine as into Netherlandish art, and perhaps into Florentine even more potently. Flemish religious art was surprisingly dependent on Stradanus's engravings. Everywhere his style and manner of composition formed the basis, as can best be seen in the work of Martin de Vos and Ambrosius Francken. The differences between Flemish and Dutch painting of the period were not greater than those between Zucchi's religious and his mythological pictures. (And when Flemish artists depicted mythological scenes, they came very close to Zucchi—for example Johan Wierix's drawing of *Diana and Calisto*, which Benesch publishes.) It is almost self-evident how important were Zucchi's frescoes in S. Silvestro, the works of other Tuscan artists active in Rome, as well as the whole circle around the Zuccari in Rome (for example the paintings in S. Lucia del Gonfalone) as stimuli for religious art in Flanders. The occasionally rather stiff, crowded compositions of Zucchi in S. Silvestro, which contain so much of Stradanus, show inevitably affinities with Martin de Vos and Joos van Winghe. Furthermore, the archaizing tendency of these Zucchi frescoes—embodying as they do Northern late-gothic elements inherent in the earlier Italian mannerist style—are to be found also in the works of most of the other Florentine painters, Ligozzi[2] and Poccetti for instance. This archaizing tendency was a feature of the period generally, and appeared most strongly in Flemish religious paintings and engravings.[3] But it played

[1] In the manner in which he handles his religious narratives, he foreshadows Martin de Vos.

[2] The Florentine court painter Ligozzi, who was born in Verona, absorbed influences from Dürer, Baldung Grien, Burgkmair and Stradanus. The most striking parallel to him was the Flemish artist George Hoefnagel, who worked at the courts of Munich and Prague. His style was late-gothic in an archaizing manner, influenced by Dürer and the marginal decorations of Netherlandish prayerbooks, yet showing already an almost scientific naturalism in his rendering of animals and plants (see E. Kris, 'Georg Hoefnagel und der wissenschaftliche Naturalismus' in *Festschrift für Julius Schlosser*, Vienna, 1927).

[3] How inevitable was the reciprocity between Flemish and Florentine religious painting becomes evident from a comparison between *The Legend of the Crucifix at Soria* by Jacopo

a dominant rôle also with Dutch artists (notably Goltzius) and with the painters at the court of Rudolph II at Prague. In the art of Flanders and of Rome, in Florence and in Holland, one meets the same archaic, predominantly Northern types, the same dependence on Dürer and Lucas van Leyden. More frequently observed have been the connections of Flemish artists with Tintoretto, particularly in the case of Martin de Vos, who was of course his pupil. The phenomenon, which we have remarked in Wtewael, namely, that when Northern artists assimilated Venetian influence the result looks like contemporary Tuscan art, is also very apparent in de Vos's work. Yet the absorption of Venetian elements by Flemish artists also produced a resemblance with the art of Verona: for example Martin de Vos reminds one of Zelotti; and Joos van Winghe, going beyond Tintoretto and the Veronese artists, recalls, rather, Zucchi. Thus the representations of *The Last Judgment* by Martin de Vos, Jacob de Backer, and Crispin van der Broeck all originated from a source which—to remain within the sixteenth century—began with Orley and Lucas van Leyden, thereafter ran through numerous variations on the theme at the hands of Floris, Stradanus, Tintoretto and Federico Zuccaro. They constituted more or less a parallel to this line of development, showing the same emphasis on linear structure of the design and of the individual figures, that was characteristic of Stradanus, of Tuscan art, and of earlier Netherlandish mannerism. In his numerous drawings Martin de Vos translated the broad and painterly qualities of Tintoretto and Palma Giovane into something nearer to Zuccaro by means of linear tautness. In his compositions the Northerner depended as much on the earlier Netherlandish mannerism—Pieter Coecke, for instance—as Zucchi did on the earlier Tuscan mannerist style. A comparison of de Vos's *Paul and Barnabas at Lystra*, 1569 (Brussels) (Plate 19), with the works of the early El Greco shows in both the

---

Coppi in S. Salvatore, Bologna (a Florentine artist influenced by Stradanus) and the *Seven Joys of Mary* by Blondeel (Tournai Cathedral) which in its composition of small narrative scenes glowing within openings in the overall architectural framework closely resembles in construction the Bologna painting. It is significant that, according to the literature, Blondeel is thought to have had close connections with Florence and especially with Pontormo. Coppi's composition, recalling contemporary Northern tomb decorations, looks like a revival of Pontormo's *Joseph in Egypt* (London, National Gallery). It is not surprising, therefore, that he found so much affinity with the mannerist artists of Antwerp and Bruges of 1520.

same heightening of Tintoretto's style, a characteristic of later Tusco-Roman mannerism.[1] In composition and in his whole artistic temperament, even in his prolific output, de Vos resembled Allori, the chief representative of Florentine religious painting, who borrowed heavily from Stradanus. On the other hand, one often meets in Flemish paintings figures taken from Allori.[2] However, this international exchange, via Florence and Rome, extended yet further: Allori and Zuccaro, Tuscan painting and the circle of artists in Rome, all these at this time influenced not only the Netherlands but Spain too. And, if many Flemish paintings of the period remind one of Spanish art[3]—a typical example is the panel by de Vos with scenes from the life of *St. Conrad* (Antwerp)—this is the result of the common, Florentine source. Thus the circle closes on this international style around 1600, whose origin was in Tuscany.

As the inter-relationship, on the formal level, of all the Northern streams of later mannerism is too obvious to permit of any such fundamental differentiation as Benesch attempts, he tries to justify his thesis by crediting the

[1] The Flemish artists, living in Venice, took over not what was most characteristic but what was most mannerist in Venetian art, i.e. mainly Tintoretto, and there only his most mannerist elements, which approached nearest to Florentine art. (Indeed the passage with the sea-gods in his large ceiling fresco *Venice as Queen of the Sea* in the Doge's Palace might be something out of Zucchi.) And to judge from our still rather imperfect knowledge of their work, those interesting painters Frederick and Gillis van Valckenborch also belong to this category of Northerners living in Venice. They developed the style of their teacher Paolo Fiammingo and transformed impressions gained from Pozzoserrato and especially from Tintoretto into something much more mannerist, thus adapting themselves well to the European style centred on Tuscany. Benesch is correct when he states that F. van Valckenborch paved the way for Callot and Bellange—two artists who are only to be understood in relationship with Tuscan art. In fact we have here the same phenomenon as at the beginning of the century in which only the superficial, and therefore perhaps more immediately striking Venetian connections have been noted and stressed, while the much more essential relationship with Tuscany has for the most part been overlooked.

[2] Stylistically related transformations of compositions by Raphael and Barocci were responsible for the similarities in Martin de Vos's and Allori's *Miraculous Draught of Fishes* (Berlin 1589, and Uffizi 1596 respectively). There is of course always the possibility that Allori may have known the version by his Flemish contemporary.

[3] Observation by Longhi.

various schools with differing intentions: in Antwerp the artist sought to edify; in Haarlem to instruct; in Prague he occupied himself with abstract formalism. Yet, in the light of the parallel Italian pioneering development, these differences prove not to be fundamental and exclusive characteristics of individual local schools, but variously emphasized, variously combined nuances of one and the same style: namely, the international later mannerism. Further to strengthen his theory of 'national' styles, Benesch advances the proposition that the art at the end of the century was chiefly concerned with spiritual problems, whereas that of the mid-century had been occupied more with formal questions. This is as artificial a distinction as that which he draws between the styles of the various countries, and serves only to underline the arbitrariness and artificiality of his view. One cannot hope to explain so simply such a complicated problem—and certainly not by using nineteenth-century criteria—though indeed we all err all too often in applying easy formulas to this difficult subject. 'Formal' and 'spiritual' elements were indeed much more closely and vitally connected in the sixteenth century than one might today suppose, for every form has its corresponding spiritual, ideological content—and the emphasis on pure content in the art-literature of the second half of the century does not alter that at all.[1] As will have been gathered from my argument, later mannerism spoke one formal language, and evolved one scheme of composition appropriate to its purpose, one objective norm with the obligatory subordination of all parts to the whole. The spiritual content of this period, i.e. at the close of the century, had thus found an adequate form by which it stands or falls: the last, and, as I like to think, for that reason, so impressive and intense, flare up of the late Middle Ages. The mannerist style persisted so long as this late-mediaeval formal spirit remained alive. It made itself felt as effectively in Holland[2] as in the Catholic countries; and under its powerful and violent aegis, in Galileo's Florence as in Holland—though in another form—the new era began.

The climax of this style, which had found its constant inspiration in

---

[1] Benesch himself stresses the esoteric preoccupation with 'form' at the Catholic courts of Prague and Munich. This connection between form and spiritual content coincides with my own argument, but contradicts Benesch's other reasoning.

[2] If one accepts this generalization, Benesch is correct when he remarks that 'mannerism survived longest in Catholic Utrecht'.

Florence, and again and again formed and reformed around Florence, was reached in the work of a Northerner: Jacques Callot.[1] Benesch regards Callot as a late mannerist strayed into the baroque age; this too is a consequence of his one-sided interpretation of the period, which fails rightly to appreciate the rôle of Italy. Although there was by Callot's time already enough baroque work in Florence to make the beginnings of an ever widening stream, it nevertheless seems exaggerated to talk in this transitional period of 'a belated mannerist', when in fact mannerism—and in Florence even more so than in the Netherlands—was experiencing still a steady, heightening development, productive, right up to Callot's time, of a series of homogenous and significant works. At the end of this evolution, Callot, one of the most important of Northern artists, arrived in Florence, and created his style based on what had gone before him: Stradanus, Zucchi, Zuccaro's paintings in the cupola of the Duomo, Poccetti, Boscoli, Salimbeni, Tempesta, Buontalenti, Parigi and Cantagallina—whilst the Northern links (de Gheyn etc.) fitted quite naturally into all this. The phenomenon of Stradanus's incorporation in the Florentine development[2] was thus repeated. Callot was the culmination of the international-Florentine line from Stradanus to Boscoli. Callot followed Boscoli in the same way that Boscoli had followed Stradanus. Thus the history of art repeated once more the pattern of the seventies, eighties and nineties. In the light of this discussion it can hardly be thought to have been by chance that mannerism survived nowhere longer than in Tuscany; and my argument is perhaps strengthened by the fact that the last great Northern mannerist should have evolved his style in large measure from Florentine beginnings.

[1] Bellange too is unthinkable without Tuscan artists like Rosso, Pontormo, Salviati, Salimbeni, etc.
[2] In the service of the Grand Duke of Tuscany between Stradanus and Callot, there were Tobias Verhaecht, the first teacher of Rubens, and Hans von Aachen.

---

*Translator's note:* Some of the more significant publications, bearing directly on the theme of this article, which was written in 1929: are Fritz Baumgart, 'Zusammenhänge der Niederländischen mit der Italienischen Malerei in der zweiten Hälfte des 16. Jahrhunderts', *Marburger Jahrbuch*, XIII, 1944; Federico Zeri, *Pittura e Controriforma*, Milan, 1957; Konrad Oberhuber, *Die stilistische Entwicklung im Werk Bartholomäus Sprangers*, (Diss.) Vienna, 1958. J. A. Gere; *Burlington Magazine*, 1960. E. K J. Reznicek, 'Studies in

# The problem of Mannerism in the Netherlands

Western Art: The Renaissance and Mannerism' in Vol. II, *Acts of the Twentieth International Congress of the History of Art*. Princeton Univ. Press, 1963. Other relevant publications are listed in the bibliography appended to the Catalogue of the exhibition *Fontainebleau e la Maniera italiana*, Naples (July-October, 1952).

# 3

# Observations on
# Girolamo da Carpi

One of the best-known and most impressive paintings at Hampton Court is the Italian sixteenth-century *Portrait of a Lady in a Green Dress* (Plate 34a). Though it has been lavishly praised and its 'handwriting' seems extremely clear, nevertheless it has presented, art-historically, a real puzzle and a satisfactory author could never be found for it. Its outstanding qualities justify a somewhat lengthy discussion of it. Belonging originally to the Duke of Mantua, it passed into Charles I's collection, where it was attributed, as so many good Italian portraits of the first half of the sixteenth century, to Raphael and Sebastiano del Piombo in turn.[1] Lately, it has been alternately assigned to Bronzino and Pontormo.[2] Only quite recently has it been considered North Italian, perhaps Ferrarese.[3] In my opinion, the picture is really Ferrarese and its author, Girolamo da Carpi (1501-*c.* 1557).

Once this name is suggested, detailed proof scarcely seems to be necessary. Whether we compare the Hampton Court picture with one of Girolamo's religious or mythological paintings or with one of his few extant portraits, the similarity is striking. The type and shape of the face, the shape of the features, the particular quality of naturalistic, slightly hard plasticity, the

[1] Horace Walpole, who in his journal recorded it as 'very good', still thought it to be by Piombo (*Walpole Society*, XVI, p. 80).

[2] The more recent Hampton Court catalogues attribute it to Bronzino. As this attribution has not proved convincing, A. Venturi ('Due ritratti inediti del Pontormo', *L'arte*, XXXIV, 1931, p. 520 ff.) assigned it to Pontormo. As such it was republished by L. Becherucci (*Manieristi toscani*, Bergamo, 1944, p. 19).

[3] B. Berenson, whose opinion is quoted in the *Catalogue of the Exhibition of the King's Pictures* (London, 1946-47, p. 82), believes the picture to be North Italian, between Lotto and Dosso Dossi, perhaps by Battista Dossi. The catalogue of the exhibition itself also considers it probably North Italian, perhaps from Cremona or Ferrara.

rather melancholy expression of the sitter, appear almost identical with, though more individualized than, those of such different subjects as the Madonna in the *Adoration of the Magi* (1531, Bologna, S. Martino) (Plate 34b), the St. Sebastian in the *Marriage of St. Catherine* (Bologna, S. Salvatore) and the *Judith* (Dresden). Furthermore, in all these pictures, as in the Hampton Court portrait, the fingers are gracefully bent and the index finger held slightly apart from the others in the same elegant manner. In colouristic effect, the cool and precise lightness of the Hampton Court portrait can be fittingly compared with that, for instance, in Girolamo's mythological pictures in Dresden; in particular, the deep, clear emerald green of the dress, which dominates the portrait, reminds one of various draperies in these paintings.[1] Naturally we must also adduce Girolamo's portraits already known. The simple portrait (1532, Pitti) of the Archbishop of Pisa, Bartolini-Salimbeni, especially, documented by Vasari, has much the same sensitive expression, a similar structure. A few portraits, too, lately attributed to Girolamo, with good insight into his art, show a close kinship to ours, though perhaps none is quite so near to Girolamo's types as this. Outstanding is the grandiose *Portrait of a Lady* in Frankfurt, which was regarded by Burckhardt as one of the most important Italian female portraits beyond the Alps,[2] and which probably represents Renée of France, Duchess of Ferrara, wife of Ercole II, noted for her Calvinist belief.[3] The severe setting of the figure within its frame, the attitude, the sharply cut, almond-shaped eyes, the elegant arrangement of the arms, hands, and fingers, the painstaking and elaborate treatment of the sumptuous green dress, all resemble our portrait. Or

[1] The same colour also occurs in the little *Adoration of the Magi* in the National Gallery, London (no. 640), there called 'Ferrarese School'. Since H. Mendelsohn, *Das Werk der Dossi* (Munich, 1914, p. 191) it is rightly ascribed, in art-historical literature, to Girolamo da Carpi. The mulberry-red background of the Hampton Court portrait is also to be found in the picture of the half-length figure of a female saint in the National Gallery, London (no. 4031), where it is attributed to Dossi. It has been correctly given to Girolamo da Carpi by R. Longhi (*Ampliamenti nell' Officina Ferrarese*, Florence, 1940, p. 37).
[2] J. Burckhardt, *Beiträge zur Kunstgeschichte von Italien*, Basel, 1898, p. 283.
[3] H. Voss, 'Girolamo da Carpi als Bildnismaler', *Städel Jahrbuch*, III-IV, 1924, pp. 97 ff. I must add, however, that this rather unexpected attribution of the famous picture does not seem to be generally accepted as there is almost no reference to it in art-historical literature.

again, just as near to it are, for instance, the portrait bust of a man in a beret, in Modena,[1] and the likeness of a richly dressed young gentleman with a porcupine, previously in the Beit Collection, London[2]: both have the same malleability in the modelling and the shading of the face, particularly of the cheek.

So much for the attribution itself. What precisely is this portrait style which the Hampton Court portrait represents and which it has taken so long to recognize and to distinguish from similar contemporary portraits in Italy, particularly in Florence?

Girolamo spent part of his youth in Bologna; here he was naturally drawn under the influence of Parmigianino, who was living in the town between 1527 and 1531 and whose altarpieces and portraits, combining new expressiveness and elegance, were there held to be the last word in fashion. There is no doubt that in its general features and sentiment our portrait shows a definite mark of Parmigianino.[3] The motif, too, so important for our picture, of the clasped hands (with gloves held between them), in which the spread, bent, and separated fingers play such a conspicuous part, is a variation on the elegance of Parmigianino. The treatment of the dress, though more graphic in character, is equally apt to recall some of Parmigianino's portraits.[4] Yet the differentiation between Parmigianino, who is more forceful and yet sometimes more ethereal, and the more matter-of-fact Girolamo should not present an insoluble problem. Even the structure, the simple dignified carriage of the Hampton Court portrait is slightly more Raphaelesque than Parmigianinesque. The cutting-off of the figure just below the hands continues a practice of Cinquecentesque classicism while Parmigianino's portraits, in mannerist fashion, usually reveal much more of the figure. In the expression of the Hampton Court picture not only is there more of everyday

[1] R. Pallucchini, *I dipinti della Galleria Estense*, Rome, 1945, p. 96, no. 192.

[2] Voss, *op. cit.*

[3] A lengthy comparison of the relation of Girolamo's portrait style, particularly in the Frankfurt picture, to Parmigianino, has been made by Voss and need not be repeated here.

[4] Longhi (*op. cit.*) even considers, and I think quite rightly, that the *Portrait of Giovanni Battista Castaldi* in Naples, attributed to Parmigianino (where not only the whole conception of the face and the expression but even the black trimming on the collar is quite close to the Hampton Court portrait) might be by Girolamo himself.

humanity than in most of Parmigianino's almost masquelike portraits, but, in particular, the personal, melancholy gaze is very different from the abstractly fixed stare which so often characterizes Parmigianino's work; for instance, the *Antea* (Plate 35b). Nor is there anything in it of the aggressive projection of the face and hands that is peculiar to Parmigianino's likenesses. Though in Girolamo's picture the face is very slightly, almost imperceptibly stonelike, its naturalism is meticulous, as it were, photographic, in comparison to the plasticity of Parmigianino's portraits, particularly his later ones, which are less real and whose chiaroscuro is less descriptive. The main colours in our portrait—the emerald green of the dress, the white of the underbodice, the mulberry-red of the background—are in intentional opposition far less than in Parmigianino; however, when compared with a Cinquecentesque classicist harmony, they do stand firmly on their own and somewhat in contrast to each other.

It is scarcely necessary to dwell upon the difference between our portrait and those of Bronzino, a difference much greater than the one we have just discussed; for, to repeat a commonplace, Bronzino's likenesses are far more impersonal and ceremonial than Parmigianino's, the individual expression of the sitter is suppressed and the plasticity assumes a marble-like character. But Girolamo's portraits are almost equally remote from those of Bronzino's master, Pontormo. True, some of Pontormo's early portraits are very expressive and 'Dürerian', but in these, too, the faces are almost geometrically constructed in oval shapes, with eyes exaggeratedly wide-open, and the decorative colours are unmistakably different. An even greater difference separates our picture from Pontormo's later portraits. The latter's hieratic *Portrait of a Lady* in Frankfurt, of about 1535 (Plate 35a), with which our portrait has repeatedly been compared,[1] already possesses all the characteristics of those

[1] Becherucci, in her account of Pontormo's career, cannot, of course, solve the problem of the Hampton Court portrait's proper chronological place. Nor does she appear to appreciate the dissimilarity, not just slightly formal but fundamental, between Pontormo's Dürerian portraits of about 1525, immediately after the Certosa frescoes, the time of the revolutionary politico-religious upheavals, and those of about 1535, that is, soon after the establishment of the Duchy and its court in 1531. She dates the Hampton Court portrait about 1530, bringing it, chronologically, together with Pontormo's *Portrait of a Lady* in Frankfurt, which is, as I believe with C. Gamba (*Il Pontormo*, Florence, 1921, fig. 47), unmistakably later, at least as late as 1535, for it presupposes the existence of a court.

unapproachable, distant beings with which Pontormo endowed his sitters at this mature phase. This can be seen in the rigidly upright carriage, the cold expression, no less than in the very hard, quasi Empire-like way in which it is painted, with its scarlet dress built up as a great abstract mass. It is not so much the pursuance of details as the less affectionate manner of doing it which separates Pontormo, at this phase, not only from Girolamo but even from Bronzino. Neither in his late nor in his early phase would Pontormo have given such vibrant, alive facial contours as those of the Hampton Court portrait, while the chiaroscuro of his faces differs from Girolamo's exactly as does Parmigianino's. However, we do not find in Pontormo Girolamo's elegant, slender Parmigianinesque hands and it is characteristic of him that the lady in the Frankfurt picture does not show her wrists. All these individual differences, small as well as great, are significant, although Girolamo represents a certain stylistic parallelism, in a general way, with contemporary tendencies in Florentine painting and, like Pontormo and Bronzino, was a portraitist of the court and of court circles. For after his return from Bologna to Ferrara he portrayed Duke Ercole II (this full-length portrait survives only in a copy, in Modena), his children (of these pictures, some of which were sent to the French court, there is no trace) and, very probably, also the Duchess (Frankfurt).

Most writers who have described the Hampton Court portrait, even those who have thought it to be by Pontormo, have been struck by its Lottesque character. Lotto stands as a kind of symbol behind any North Italian portrait of any quality which has an individualized and soft expression. He renders his sitters intently regarding the spectator and in a more individualized manner than anyone else of his time in northern Italy, not excluding even the young Titian. Girolamo knew Titian well personally, and undoubtedly the works he made for Ferrara; by order of Ercole II he had even copied, for the king of France, his very grandseigneurial *Portrait of Alfonso I*.[1] Also, one can take it almost for granted that he visited Venice from nearby Ferrara and possibly saw portraits by Lotto there. But it would be going too far to say that he was greatly under Venetian influence. We come nearer to the

---

[1] In his list of Girolamo's paintings (*Italian Pictures of the Renaissance*, Oxford, 1932), Berenson includes the version in the Pitti, as Girolamo's copy after Titian.

Lottesque character of the Hampton Court portrait if we turn to Dosso Dossi,[1] the outstanding artistic personality of the Ferrarese court milieu in the years before Girolamo's activity. Remnants of the Giorgione-Ariosto spirit never entirely die out in Dosso,[2] at least not in his portraits or portrait-like half-figures. At the same time, original even when eclectic, Dosso, in using Central Italian models, altered the expression through accentuation.[3] So Girolamo's Hampton Court portrait quite naturally has a certain resemblance in expression to some Dosso portraits, though in Dosso not only is the treatment of the face far more summarized in a Cinquecento classicist sense, but even the vivification of the expression is general rather than penetrating in character. Incidentally, Girolamo's particular, darkish green, purer than that of Parmigianino, harder than that of the Cinquecento classicist style, approaches the emerald green of Dosso, especially of the late Dosso.

But still another North Italian influence, Emilian, was exerted on Girolamo, expediting the spiritualization of expression. We know from his own account, given by Vasari, how persistently in his youth he followed up and copied Correggio's works in Bologna, Modena, and Parma.[4] In fact, the Hampton Court portrait (as well as numerous other pictures by Girolamo), though less charged with emotion, has undoubtedly something from Correggio in the sentiment, precisely in the unusually deep look in the eyes. If we wish to define Girolamo's portrait style, with reference to the expression, in

[1] It is no accident that even quite lately Dosso's and Lotto's portraits have been confused.
[2] Giorgionism, Ariosto, and the early Dosso all have certain, but only certain, features in common. A precise analysis of their resemblances and diversities will be possible only when the whole social and cultural environment of these North Italian artistic and literary phenomena has been examined. Dosso, the Ferrarese court artist, brings out of Giorgione's much more tranquil classicism the fantastic elements and strongly accentuates them. He does this, in particular, when illustrating Ariosto, for—although it is at the moment unfashionable to say so—Dosso and Ariosto are very closely related in spirit. On the relation between the two, see J. Schlosser, 'Ferrara' and 'Der Weltenmaler Zeus' (both essays in the volume *Präludien*, Berlin, 1927, pp. 160 ff. and pp. 296 ff.).
[3] For instance, by the well-known transformation of Raphael's *Giuliano de' Medici*, he produced a bold, almost brutal portrait of a warrior with glittering eyes (Uffizi).
[4] Voss thought that Girolamo stressed his links with Correggio's manner only to establish his ancestry from among the great painters and divert attention from his imitation of Parmigianino, his younger contemporary. But there is more in it than this (see also note 21).

terms of direct influences, it is really Dosso and Correggio who take Lotto's place in it.

Of course influences alone cannot explain an artist or do justice to him. Yet, from all these similarities, parallels, and differences, we have perhaps succeeded in circumscribing and defining Girolamo's particular place within Italian portrait painting, between classicism and mannerism: a combination of dignity and unobtrusive sentiment, of hard, though not unpliable, plasticity and naturalism. Before the Hampton Court portrait we can but underwrite Vasari's praise of Girolamo as a portraitist and recognize that his likenesses belong among the most outstanding in North Italy.

How is the Hampton Court portrait related to Girolamo's *œuvre*, to his general style and development?

Girolamo's art grew, organically, from that of his native town, Ferrara. There he spent most of his days, apart from two long absences: one, in his early phase, in Bologna and one, at a later date, in Rome. As Ferrarese art was at that time a kind of junction point between North and Central Italian art, the geographical direction of these two sojourns in itself indicates the tendency of Girolamo's style. In fact his figure style, although Emilian in character, is more akin to that of Parma and Bologna than Venice. It tends, in essence, southward toward Central Italian painting. In this regard it differs from Dosso Dossi (especially early Dosso) and resembles more closely his Raphaelesque or classicizing contemporaries in Ferrara such as Battista Dossi, Garofalo, and even Ortolano.

We know little of Girolamo's earliest works in Ferrara. Apparently he began in Garofalo's atelier, where he was working in 1520,[1] and seems to have collaborated with him on a few works, perhaps even after becoming independent, when he amplified his master's classicism, which was growing rather primitive and sterile.[2] The exact duration of Girolamo's ensuing stay

[1] We can, with some safety, trust the document published by Baruffaldi, as it corroborates the stylistic evidence.

[2] Girolamo's early Ferrarese style is a problem in itself. A more thorough analysis than has hitherto been made of certain pictures executed by Garofalo, from about 1520 on, in more or less loose collaboration with others, would yield some, though not very significant, results. A. Serafini's book, *Girolamo da Carpi*, Rome, 1915, which deals, but quite inadequately, with this problem, shows no understanding of Girolamo's style and de-

in Bologna is not quite certain; perhaps he remained there (though certainly with interruptions)[1] for as long as from the mid-twenties to the mid-thirties. However conservative the native Bolognese artists themselves may have been, his sojourn there represented for Girolamo his sight of the world, his most intense time for learning, his outpost for viewing modern art, the great masterpieces located in Bologna and the neighbouring towns. One can, without difficulty, imagine the reaction of this sensitive, susceptible Ferrarese, who knew only the art of his native town and who did not go to the source itself, Rome. Instead he went to Bologna, where different artistic tendencies intruded upon him and where he was exposed to many and varied, rather isolated artistic shocks. That is why, though his general tendency remained classicist, the works he produced here—apart from portraits, all religious themes—show wide divergences from each other. Further, it must sometimes have been difficult for him to assimilate his various impressions into an harmonious whole and his pictures of this period are usually less all of a piece than his later ones; when he specifically tried to introduce movement into the composition, as in the two *Adorations* of S. Martino (Plate 32a) and Modena, the effect is somewhat spasmodic and disruptive. The influences in his Bolognese works, though varying in degree, can easily be identified and are well known.[2] When Girolamo is classicist, his classicism is more monumental, more super-Raphaelesque than that of the Dossi. When he displays mannerist features, they are more modern, up-to-date than those of the

---

[1] These interruptions would have included a stay in Ferrara in 1530 since there is documentary evidence of his frescoes in S. Francesco in that year. In these frescoes the types of female saints in the medallions show the influence of Parmigianino, which Girolamo must already have undergone in Bologna.

[2] Voss has stressed the influence of Parmigianino, Longhi that of Raphael and Giulio Romano.

---

velopment and its catalogue of his works is extremely unreliable. However, as it gathers together many documents concerning Girolamo and enumerates most of the works then attributed to him, it serves as a starting point for future critical study. Voss (*op. cit.*), however, who deals mainly with the portraits, and Longhi (*op. cit.*), who also deals with only a few of Girolamo's works, chiefly those done in Bologna, have defined many features of his style. Incidentally, illustrations of most of Girolamo's works mentioned but not reproduced here can be found either in the books by Serafini and Longhi or in Voss's article.

Dossi. In attitude, expression, draperies, they echo Giulio Romano and, even in complete parts of the figure pattern, Parmigianino's *Madonna of S. Margherita* and *Madonna della Rosa*, painted at the end of the twenties in Bologna. Beside the influence of Raphael and Parmigianino appears also that of Correggio, on whom Girolamo himself is reported to have laid such stress[1]; we know, in fact, that he made copies after all three artists. Compared to the works of Parmigianino, Girolamo's religious compositions are more this-worldly and human in their conception, drier in their handling, less sweepingly rhythmic, while Parmigianino's extremely spiritual and spirited types, though they have deeply penetrated Girolamo's art, are never taken over literally. Girolamo individualizes his types more than any of the artists who influenced him, not only more than Dosso, but more even than Parmigianino. To some extent he continues the sharply chiselled, classicist types of Ferrarese painting, of Garofalo and in particular of Ortolano[2]; but here again

[1] Correggio's *Adoration of the Magi* in the Brera probably influenced Girolamo's picture of the same theme in Modena and that of Dosso in the National Gallery, London, simultaneously. The baroque atmosphere which Longhi rightly perceives in Girolamo's Muzzarelli altarpiece, *Madonna in Clouds with Angels* (painted for S. Francesco in Ferrara, now in the Kress Collection, New York), goes beyond Raphael, whom Longhi mentions as Girolamo's model for this picture: it derives even more from Correggio, and in particular, as far as the upper part of the picture is concerned, from the *Madonna of St. Sebastian*, one of the numerous works by Correggio which Girolamo copied. Precisely because of his strong classicism, Girolamo could even occasionally take a step toward baroque; for baroque is a less irrational sequence to a rational, realistic classicism than is mannerism. That the elements of the baroque in the sixteenth century (not to speak of Venice) were contained in the Stanza d'Eliodoro and in Correggio is by now generally realized.

[2] Even types of the kings from Niccolo Pisano's altarpiece of the *Madonna with the Quattro Incoronati* (1520, Ellis Collection, Worcester, Mass.) seem to reappear in Girolamo. R. Longhi (*Officina Ferrarese*, Rome, 1934, pp. 125 ff.) is, to some extent, right in describing the art of Niccolo Pisano and Ortolano as classicism for the people; their large altarpieces have certainly some popular features (though Niccolo painted a good deal, and even secular subjects, for the Ferrarese court, too). It would be interesting to analyse the circumstances, within a systematic, stylistic (not merely formalistic) history of Ferrarese painting, under which this somewhat popular ecclesiastical, classicist art was absorbed in the next generation, largely through Girolamo da Carpi, however transformed, and on a new, somewhat mannerist level, into an art of an increasingly elegant, courtly character. This continuation and transformation in a sense parallels a contemporary phenomenon

he gives these types, which are very schematic in spite of their seeming realism, a far more personal appearance. Indeed, the frequently recurring, almost portrait-like faces, so reminiscent of the conception of the Hampton Court picture, are a main characteristic, Ferrarese or not, of Girolamo's altar-pieces.

If I mention the impression which Peruzzi's cartoon in sepia of the *Adoration of the Magi* (Plate 33b) must have made on Girolamo, I do not wish merely to add yet another 'influence' to an already long list. This work was executed in Bologna in 1522 by order of Count Giovanni Maria Bentivoglio in whose house it hung (now in the National Gallery, London).[1] With its amazing display of archaeological-architectural motifs and reminiscences of antique sculpture, it stands for an 'antiquizing' statuesque tendency which not only is apparent in Girolamo's Bolognese works but which will, from now on, become one of the conspicuous features of his art.[2] Peruzzi's importance, particularly in Rome for Giulio Romano and others, was exceedingly great, as early as the second decade of the century, in the evolution of an 'archaeo-logical', somewhat tempered, mannerism.[3] Thus, in a general way, the

---

[1] Leaving aside Siena and Rome (Logge), as seen from Florence, the line of origin of Peruzzi's cartoon passes from Leonardo's *Adoration of the Magi* through the late Botticelli's composition of the same theme: the potentially mannerist features of the Quattrocento lead up to mannerism. Correggio's early Brera *Adoration*, which also could almost be counted among the ancestors of Peruzzi's cartoon, forms the North Italian parallel to the later stage of this development. It is perhaps not surprising to find concurrences even between Peruzzi's cartoon and Liberale da Verona's *Adoration* in the Cathedral of Verona; this last-named picture, in accentuated Quattrocento gothic, just as Correggio's early mannerist one, derives, fundamentally, from Mantegna's *Adoration* in the Uffizi.

[2] In Girolamo's native city of Ferrara, a sculpturesque tradition already existed in the story-telling friezes in grisaille, imitating ancient bas-reliefs, on the altarpieces of Roberti, Mazzolino (in Bologna: Aspertini), etc. But the original impression of this Ferrarese tradition on Girolamo (who, of course, also knew many of Mantegna's works) must have been much strengthened and widened by Peruzzi's modern, monumental composition.

[3] Indeed, Peruzzi's great initial importance for early mannerism, though scarcely noticed, cannot be over-emphasized. (The best analyses of his style are to be found in G. Gombosi, 'Sodomas und Peruzzis Deckenmalereien der Stanza della Segnatura', *Jahrbuch für Kunst-*

---

in Florence, namely, the merging of stylistic features of the late Fra Bartolommeo and his popular monastic followers, like Fra Paolino and Sogliani, into the triumphant courtly mannerism.

tendency of his influence on Girolamo was similar to that of Giulio Romano. But this cartoon, an outpost of Roman mannerism, which Girolamo had before his eyes in Bologna, could not have failed to impress him; moreover, Peruzzi's style, nearer to classicism and more moderate than Giulio's, would

---

*wissenschaft*, 1930, pp. 14 ff., and in the article on Peruzzi by F. Metz in Thieme-Becker, XXVI, 1932.) He was thirteen years older than Pontormo and Rosso and his works, done in Rome between *c.* 1511 and 1516, though not of the subjective adventurousness and extremist tendency of these Florentine artists (whose style, in the decisive years of mannerism, was expressed mostly in religious painting) greatly precede them in point of time. I refer not only to the various extensive frescoes with mythological themes in the Farnesina, but, for instance, to the *Presentation of the Virgin in the Temple* in S. Maria della Pace. Nothing, of course, would give a more incorrect idea of the artistic situation in Rome than to imagine the actual relation between Raphael and Peruzzi to have been competitive, even stylistically. They can be considered to have collaborated not only on the ceiling of the Stanza della Segnatura but even more so on that of the Stanza d'Eliodoro, for which there exist drawings by Raphael as well as by Peruzzi, while the execution is, I believe, by the latter. Furthermore, it is well known that the assistants of the late Raphael moved decidedly in a mannerist direction even in the master's lifetime. (Perhaps one might mention here the potentially mannerist elements in some of the most classicist sculptures in Rome, namely, Andrea Sansovino's statues, such as the *Temperanza* on the Basso grave in S. Maria del Popolo, from as early as 1507.) Peruzzi developed with astonishing rapidity, even within the Farnesina, from the earlier frieze in the Stanza del Fregio, still based mainly on Pollaiuolo and Pinturicchio, to the later one in the Sala delle Colonne, which transforms Raphael and Michelangelo into a more or less full mannerism. His works did much to bring about the mannerist evolution in Rome, particularly among Raphael's pupils and also in his younger fellow countryman, Beccafumi. It was his mannerist archaeological interest in the antique and his equally experimental, somewhat abstract conception of space and of the human body in relation to it which made his works of such decisive significance in the Eternal City. These features influenced a long line of artists of similar tendency, beginning with Giulio Romano's and Polidoro's frescoes in the Villa Lante, depicting scenes from Roman history, and not even terminating with Pirro Ligorio. (On Giulio Romano's borrowings from Peruzzi, see H. Dollmayr, 'Giulio Romano und das classische Altertum', *Jahrbuch der Kunsthistorischen Sammlungen in Wien*, XXII, 1901, p. 216.) In his particular view of the antique, Peruzzi is anticipated somewhat by Aspertini and Ripanda but the genealogical line of his whole style, going back to Sienese and Florentine Quattrocento gothic (and even, if we wish, to Roman Quattrocento sculpture), is much longer.

So many continuous, irrational tendencies run through art from the fifteenth to the sixteenth centuries, that it is almost useless to inquire at exactly what moment mannerism 'began' in Rome, or elsewhere, for that matter. Around, and soon after, 1500 such an

have appealed to his taste no less than that of Giulio, who was indeed be-coming, in the frescoes at Mantua, increasingly heavy, dramatic, almost menacing in character. Even so, Peruzzi's cartoon, in itself not in the least harmonious, must have appeared to him somewhat violent. Its influence on him has not previously been remarked upon, perhaps[1] because of Girolamo's habit of toning down extreme features of a work of art that otherwise inter-ested him. In his *Adoration of the Magi* in Modena the modification amounts to a kind of fusion with traits of Begarelli's *Adoration of the Shepherds* in Modena Cathedral (1527)—a group Girolamo knew and which, with its

---

[1] Nor, for that matter, has its influence on Alfonso Lombardi's relief of the *Adoration of the Magi*, on the Arca di S. Domenico in Bologna (Bologna, S. Domenico), made in 1533 (Plate 36a) when Girolamo was working in Bologna, been noticed. The relief is little more than a free, compressed copy of the principal lower part of the cartoon: the general grouping of the figures, with the two large columns behind them, is similar. The details are also similar: the central group of the Madonna with the kneeling king, the spectators on the left, pressed diagonally toward the column, on the right the rearing horse and the group around the open treasure casket, with the vase beside it, while even the excitedly braying ass recurs. That the most prominent sculptor of the town should have made these quite obvious borrowings in such an all-important place as the Arca di San Domenico speaks for the great influence of the cartoon in Bologna itself. In fact, there exist numerous copies of it, done in Bologna: one, for instance, by Girolamo da Treviso (London, National Gallery), another (previously in Dudley House) attributed to Prospero Fontana, a third (previously owned by the Rizzardi family, Bologna) to Cesi. One can see this influence as late as in Passarotti's *Adoration of the Magi* (Bologna, Palazzo Arcivescovile), and even Agostino Carracci engraved Peruzzi's composition in 1579.

---

enormous bulk of naturalistic procedure had been accumulated that artists who broke through this material and used it to express themselves in a fundamentally unrealistic or even anti-realistic way, can certainly be referred to, at least for the sake of convenience in terminology, under the common denominator, 'mannerists'. I must also add that, though I have stressed Peruzzi's importance among these latter artists, I have no intention of taking part in the presentation, now in vogue, of the 'earliest of early' mannerist artists and works of art in the *very* first years of the sixteenth century. Mannerism need not be approached in terms of a race between individual artists. The interweaving and the occasional simultaneity of the two main currents with their far-reaching ancestry—the classicist and naturalist, and the anti-classical or mannerist—and the prevalence of one or other of them at any given period, will be entirely understood when treated together, fully and very concretely, in their wider social setting, and will no longer hang in an airless vacuum of isolated individual artists.

classicist and Correggiesque character, offers certain stylistic parallels to him.[1] Despite this softening, individual motifs reveal clearly the marks of the Peruzzi cartoon, although for a correct notion of Girolamo's borrowings one must consider some four of his works together: the *Adorations of the Magi* in Modena, in S. Martino in Bologna, and in the National Gallery in London, and the *Marriage of St. Catherine* in S. Salvatore in Bologna. The heads of St. Joseph in Modena and in S. Martino, for instance, so remindful of the antique, are near copies of Peruzzi's God the Father[2]; the group in S. Martino with inclined heads and shoulders, bending forward and pressed tightly together behind the Ethiopian king, appears to derive from Peruzzi's attentive on-lookers. The Christ Child in Girolamo's altarpiece in S. Salvatore seems also to come from the same cartoon. The half-nude man in the Modena *Adoration*, unusual for this subject, seen backview and in profile[3] (and placed on a high plane, like the onlookers in the cartoon) recalls Peruzzi's pseudo-nude, 'antiquizing' figures in their skin-tight garments, in particular the warrior, likewise shown backview and in profile, standing behind the Madonna.[4] The arrangement of the figures lifting up the gifts on the right-hand side of Giro-lamo's small *Adoration of the Magi* in London exhibits the unmistakable in-fluence of the corresponding group around the treasure chest in Peruzzi's cartoon.[5] What, indeed, is significant, beyond the individual reminis-

[1] On account of a similar nearness of its figures to the ground, Girolamo's Modena picture can well be compared, in its general impression, with Begarelli's composition, where most of the figures are also bent or stooping.

[2] At the same time, these heads, both in Girolamo and Peruzzi, are further evolutions of that of Raphael's *Ezekiel*.

[3] Perhaps one should also mention, as a general source for this figure, Begarelli's numer-ous half-nude shepherds.

[4] The man holding the rearing horse (Group of Montecavallo), seems to have some bear-ing upon Girolamo's (otherwise Dossesque) negro king at Modena, a figure for which the shepherd swiftly approaching with outstretched arms and legs, on the extreme right of Begarelli's *Adoration*, may also have been important. I realize I have cited a large number of Peruzzi and Begarelli figures but I believe them to have been quite significant in the formation of Girolamo's style, even though the general resemblances are frequently not the result of direct, blatant borrowing.

[5] This same picture of Girolamo, which demonstrates, stylistically, the influence of Giulio Romano, has also a touch of Mazzolino's early mannerism. The fact that Peruzzi (as we know from Lamo's *Graticola di Bologna*) praised an altarpiece of Mazzolino at the time of his stay in Bologna, fits into this context.

cences,[1] is the similarity of the general 'archaeological' tendency in both artists. The antique draperies in Peruzzi's cartoon, which with emphasized flutings cling very closely to the body, or form round it a detached, metallic covering, or float in loose ornamentality, must have stimulated Girolamo's construction of cold, statuesque, monumental figures, resembling ancient sculptures. This archaeological tendency, when pronounced, is close to mannerism, and when slight, is not far from classicism. The former occurs, to some extent, in the intense and original Peruzzi, in whose frequently crowded works this tendency is coupled with a somewhat ornamental, abstract treatment of the figure. The latter occurs in the much quieter, more naturalist pictures of Girolamo, who in his early period only rarely employs Peruzzian traits and only in a moderate and diluted form. His reliance on a Raphaelesque classicism, however transformed, prevents him from accepting, with consistent readiness, pronouncedly mannerist innovations; yet Peruzzian features, however modified at first, remain as a spiritual 'antiquarian' attitude, a distinct and, as we shall later see, increasingly distinct note in his art.

From Bologna Girolamo returned to Ferrara where he is mentioned for the first time in the account books of the Ducal household in 1537—for his work for the court began at this time. It was in Ferrara, during the main phase of his career, that Girolamo evolved, at least in a few outstanding works, that characteristic, settled style with which one usually associates his name. In his native milieu he slowly absorbed the strong impressions he had received in Bologna, Parma, Mantua. The altarpiece of the *Pentecost* (Plate 38b), painted in Ferrara for the church of S. Francesco in Rovigo probably soon after his return, already shows the artist's new ability to clarify the composition of this ecstatic subject by harmonizing even agitated attitudes. In other respects, with its one or two astonishingly individualized apostles and some rather Parmigianinesque types, the painting approaches stylistically Girolamo's last large Bolognese work, the altarpiece in S. Salvatore.[2] Most

---

[1] It is characteristic that even at the present time some confusion persists between works of Girolamo da Carpi and Girolamo da Treviso, who carried out Peruzzi's cartoon in painting and also shows its influence in other of his works.

[2] We also have external evidence for the dating of the *Pentecost*. Mr. P. Pouncey of the Department of Prints, British Museum, has recognized a pen drawing of *Christ and Thomas*, in the Albertina, there attributed to Polidoro (*Die Zeichnungen der toskanischen,*

of Girolamo's work in Ferrara was done for the court, although he frequently executed paintings for monasteries and churches, for aristocrats and humanists. In embellishing the various residences of the Este family he often had to collaborate with other artists who were inferior to him and who differed from him in their tendencies. We shall first examine his style in the few independent pictures which survive. Vasari remarks that he was at his best in portraits and single-figure pictures, not in scenes of a grandiose character. The latter judgment is a typically Florentine-academic exaggeration but it holds a degree of truth in as much as Girolamo's style, too concentrated and quiet for Vasari's taste, came fully into its own in compositions with relatively few figures. The Gallery of Dresden offers a unique occasion to appreciate Girolamo's place among the other Ferrarese sixteenth-century artists[1]: only Dosso Dossi and Girolamo, each in his way, stand out strikingly by force of originality and quality, while the mythological pictures of Garofalo and Battista Dosso (leaving aside the latter's amusing 'Flemish' capricci)[2] appear rather

[1] When their rule came to an end in Ferrara (1597), the Este took their choicest pictures to Modena, from whence the hundred best were sold to the King of Saxony (1745).

[2] Battista Dosso included in his compositions not only fantastic 'northern' landscapes but even Bosch-like figures, as in the *Dream*, Dresden. (Concerning the possibility of a concrete Flemish model for this picture, perhaps Bosch himself, see G. de Tervarent, 'Instances of Flemish Influences in Italian Art', *Burlington Magazine*, LXXXV, 1944, pp. 290 ff.) These Flemish capricci form one of the most interesting features of Ferrarese court art. They are a very characteristic item within a scholarly type of painting, revealing how deeply irrational elements (continuing, if one may so put it, the 'pre-Bosch' tendencies of Tura's Quattrocento gothic) were contained in this whole courtly culture. (The *Dream* itself is built up from passages of Statius's *Thebais* and Lucian's *Verae Historiae*; see Tervarent, *op. cit.*) Nothing could better illustrate the spirit of the Ariosto-Dosso trend than that Battista Dosso, who, though much more of a classicist than his brother, should have

---

*umbrischen und römischen Schulen*, Vienna, 1932, no. 168), as being a preparatory sketch for the monochrome composition on the marble front of the dais, on which Mary and two of the apostles are seated, in the Rovigo picture. The drawing, however, with its rather crudely expressive faces is not by Girolamo himself but by the Bolognese, Biagio Pupini, with whom Girolamo had worked in Bologna and for whom he had apparently procured an invitation to Ferrara, since Biagio is mentioned in 1537, together with Girolamo, in the accounts for paintings in the Ducal Palace of Belriguardo. Thus Biagio must have collaborated with Girolamo on the *Pentecost*, which would be dated about or soon after 1537 (there is no indication that Biagio stayed longer in Ferrara).

second-rate, provincial, and unoriginal. True, it was again with Garofalo and still more with Battista Dossi that (as the account books testify) Girolamo usually worked, sometimes even on the same picture.[1] True also, his mythological pictures tend to continue in a broad classicizing sense the late ones of Battista. But, with his new artistic experiences, he grew much beyond this backward, native Ferrarese level and quality. What in Battista is a reliance on Raphaelesque models and a heavy, rather dull imitation of them, becomes, in Girolamo, elegant, poised, naturalistic in a sculpturesque sense, often trans-

---

[1] Among Battista's compositions, painted for the Duke, Girolamo's share seems to have been particularly large in the *St. George* in Dresden and in the *Hora* of the same gallery (see note 3, p. 135).

---

interpreted Ariosto in 'Flemish' taste, with 'Flemish' landscapes (*Orlando's Fight with Rodomonte*, Collection Bondy, Vienna). This is just a variation on the procedure of his brother, who illustrated Ariosto by transforming Giorgione into something bizarre. This 'Flemish' spirit in Ferrara was not restricted to Battista, for later on (c. 1545-55) a certain number of genuine Flemish artists were engaged at the court; they not only worked mainly on cartoons for tapestries (landscapes; grotesques for the borders), woven by Flemish artisans on the spot, but did some painting as well. Apparently the fanciful, panoramic vista of the landscapes commended these artists to the court, where a taste existed for imported Flemish landscapes of the Patinir type (probably many of the pictures in the Borghese Gallery, by Patinir and by his followers, came from the Ducal Castle in Ferrara). These artists collaborated in cartoons, just as in paintings, with the Ferrarese, with Battista Dossi himself in his last years, thus creating in Ferrara an atmosphere of constant interaction. The Flemish artists enumerated in G. Gruyer (*L'art ferrarais*, Paris, 1897, II, pp. 467 ff.) still await an art-historical resurrection—a work of research which, although limited by the surviving material, would throw a vivid light on the intellectual life of the Ferrarese court. Even Girolamo da Carpi, when composing grotesques for tapestries in 1548, apparently caught something of this Flemish sense for the fantastic. The landscape with Patinir motifs and figures in Bosch style, in the Borghese Gallery, which hangs as a companion piece to the well-known one by Abbate, has been attributed by C. Gamba ('Un ritratto e un paesaggio di Nicolo dell'Abbate', *Cronache d'arte*, I, 1924, p. 84) to Girolamo da Carpi. Although I believe this assumption to be wrong and that the picture is by a follower of Battista Dossi (perhaps even by a Fleming, working in Ferrara), that such an attribution could plausibly have arisen shows the complexity of this sophisticated court art. It is, perhaps, this streak of Flemish fantasy in Girolamo which later revived in him, in another, more erudite form in the bizarre-mannerist aspects of some of his studies after the antique in Rome. We shall speak about these sketches later in detail.

fused with shy, tender sentiment, a very personal mixture of classicism and mannerism.[1] Before we discuss this court style, which has never been thoroughly studied, let us first consider the court itself.

After the death of the Dossi brothers, Girolamo slowly came to be Ercole II's leading artist, just as he was that of the Duke's brother, Cardinal Ippolito, for whom he worked most of his life. The taste of Ercole II (1534-1559) and of his brother was bound to differ from that of their father, Alfonso I (1505-1534), just as the ages in which they lived differed.[2] Alfonso was persistently demanding (with varying success) Bacchanals, Triumphs of Bacchus, and Ledas, not only from his favourite artist, Titian, but also from many others, whether—to mention only the non-Ferrarese artists—Pellegrino da San Daniele or Giovanni Bellini, Raphael or Michelangelo. Giovanni Bellini's and Titian's *Bacchanals* form the original nucleus and starting point for all the later mythological pictures executed for the Duke. Yet in character alone Ercole was very different from his father, the passionate and frankly sensuous art collector. He was better educated than Alfonso, and showed a keen interest in literature; he esteemed art, above all, as a necessary means for upholding the splendour of his court and for the glorification of his person, and had a liking for scholarly, allegorical compositions which reflected his personal qualities and his personal politics. For general conditions themselves had completely changed in Ferrara. Alfonso still had power enough to pursue an energetic military policy to protect the independence of his state: Titian's *Tribute Money* with its pictorial representation of 'Quod est Caesaris Caesari, quod est Dei Deo' illustrates Alfonso's political device, directed against the aggression of the Pope. But Ercole, living at a time when Ferrara was already much weakened, was obliged to keep up the most evasive attitude and to juggle, with the greatest caution, between the great powers; his device was Patience. No wonder he inclined toward subtle, esoteric, political allegories.[3]

[1] It seems to me quite possible that the cold, tranquil, elegant classicism of mythological sculpture by Antonio Lombardi, who very probably spent the entire last ten years of his life (*c.* 1506-16) in Ferrara, in the service of the court, has left its mark on Girolamo.

[2] Dosso Dossi died in 1542, Battista in 1548. Battista was also the artist of the 'little court' of Alfonso I's mistress, Laura de' Dianti, and of their son.

[3] See also R. Wittkower, 'Patience and Chance: Story of a Political Emblem', *Journal of the Warburg Institute*, I, 1937-38, pp. 172 ff.

Admittedly, something scholarly had not infrequently crept into Alfonso's commissions too, but the resulting pictures were usually very different stylistically because the spirit of the two reigns was very different. When Alfonso ordered Bacchanals from Titian, based on the famous descriptions of paintings by Philostratus (*Feast of Cupids, The Andrii*), surely on the advice of Calcagnini, his favourite, very learned humanist, the two resulting well-known compositions, now in the Prado,[1] though keeping close to the antique text, are full of unbridled energy and joy of life. Very different was the outcome when Ercole II, who apparently was not particularly partial to Titian, ordered from Battista Dossi, probably again on the advice of Calcagnini, a reconstruction of another picture described by Philostratus, *Hercules among the Pigmies*. This painting, of course, was nothing less than a self-glorification, under cover of his antique namesake—the motive frequently behind the Duke's numerous orders for Hercules stories. Battista's picture (now in Graz) which also follows the antique text,[2] preserves a dry, hard pseudo-classicism too inflexible to permit the solution of the contrast between the large nude body—the head of Hercules is a portrait of the Duke—and the many small ones. Although the Dosso brothers frequently collaborated on the same work, it is perhaps permissible to read into the accounts that Alfonso preferred Dosso Dossi (whose gay nature appealed to him too); Battista (who lived on into the new Duke's reign) was preferred by Ercole.

Ercole also sometimes commissioned erotic themes and Bacchanals but these were not now so exclusively fashionable and, at any rate, were conceived in a very different, much cooler, spirit. When, for instance, the old Garofalo painted the *Triumph of Bacchus* (Dresden) (Plate 40), he took as his

---

[1] See F. Wickhoff, 'Die Andrier des Philostrat von Tizian', *Jahrbuch der preussischen Kunstsammlungen*, XXIII, 1903, pp. 118 ff. and R. Forster, 'Philostrats Gemälde in der Renaissance', *Jahrbuch der preussischen Kunstsammlungen*, XXV, 1904, pp. 15 ff. On models for Titian's figures in antique statuary, see Th. Hetzer, 'Studien über Tizians Stil', *Jahrbuch für Kunstwissenschaft*, I, 1923, pp. 232 ff.

[2] See Forster, *op. cit.* and Schlosser, 'Weltenmaler Zeus', (*loc. cit.*). For curiosity's sake, I should like to mention that Hogarth, when representing Gulliver among the Lilliputians ('Hogarth and his Borrowings', *Art Bulletin*, XXIX, 1947, p. 38), unconsciously reconstructed the same picture, described by Philostratus, since Swift, in his book, had been stimulated by this antique text.

principal model a drawing by Raphael[1] and perhaps motifs from other sources.[2] These give the impression that they have been combined and heaped up—apparently the more, the better—and frozen.[3] We know from Vasari that it was precisely this picture which the Duke took great pride in showing to Paul III during a visit of the latter to Ferrara. Artists outside Ferrara whom Ercole II seems to have preferred even to his own court artists, the Dossi brothers, were mannerists: Pordenone and Giulio Romano. Soon after the erection of his tapestry factory in 1536, an artistic measure very typical of him,[4] he tried hard to persuade Pordenone, the outstanding mannerist artist of North Italy,[5] to come to Ferrara to design cartoons with

[1] This composition, much looser and freer than that of Garofalo, is preserved in a drawing of the Raphael workshop, in the Albertina. According to H. Tietze ('Annibale Carraccis Galerie im Palazzo Farnese', *Jahrbuch der Kunsthistorischen Sammlungen in Wien*, XXVI, 1906, p. 115) it is the same drawing which Raphael had sent to Alfonso I.

[2] Logge, Titian, antique sculptures, etc. One of the chief antique models for the various representations of the Triumph of Bacchus seems to have been a sarcophagus with this theme, now in the Palazzo Rospiglioso. At that time it was repeatedly copied, even as early as Aspertini (see C. Robert, 'Über ein dem Michelangelo zugeschriebenes Skizzenbuch auf Schloss Wolfegg', *Mitteilungen des Deutschen Archeologischen Instituts, Römische Abteilung*, XVI, 1901, p. 232). Another sarcophagus with the same theme, perhaps that which was at one time in the Villa Albani, was copied by Girolamo da Carpi himself (Uffizi). In Garofalo's picture inevitably occur motifs similar to ones on both these sarcophagi.

[3] Garofalo's ambition to amass as many 'interesting' motifs as possible in his mythological pictures had now reached a stage much beyond that of 1526 when, in his *Sacrifice to Ceres* (London, National Gallery), basing himself on, though vivifying, a woodcut of the *Hypnerotomachia Poliphili*, he restricted himself to six figures lined up in a row. A still earlier stage is represented by his *Mars and Venus* in a large landscape (Dresden), a picture which still has upon it a glow of Giorgione.

[4] Cosimo de' Medici, Grand Duke of Tuscany, Ercole's rival in every respect, wishing to imitate him, enticed one of the Flemish founders of the Ferrarese tapestry works to start one in Florence. The most important of the tapestries made for Ercole seem to have been a series with *Stories of Hercules* and one of the *Metamorphoses*, the cartoons for which were designed mainly by Ferrarese artists, in particular, by Battista Dossi. Girolamo, too, worked on tapestry cartoons but not intensely.

[5] Pordenone presents a good example of a mannerist who, from provincial, almost popular beginnings, developed an elegant, as it were, aristocratic, art. Pordenone, in his late years in Venice, was overwhelmed, even at Titian's expense, with official and private orders (see G. Fiocco, *Giovanni Antonio Pordenone*, Udine, 1939, pp. 89 ff.) and it was certainly

themes from the Odyssey and the Labours of Hercules. He finally succeeded in 1538, and Pordenone died there soon afterward. Like his brother, Cardinal Ippolito, the Duke gave various commissions for cartoons for tapestries, from mythology and Roman history (*Triumphs of Scipio*), to Giulio Romano, in spite of his being a servant of the court of Mantua.[1] Within the range of his frigid taste,[2] he must have included Girolamo's intellectual, elegant, polished art, with its both classicist and mannerist aspects, though he was perhaps less interested than his brother in our artist. Cardinal Ippolito, not only a greater devotee of art than Ercole but, in fact, one of the greatest patrons of his time, had the characteristic taste of a cultured expert about the middle of the century. He entrusted the erection of his house in Fontainebleau to Serlio, the Villa d'Este in Tivoli to Pirro Ligorio, and employed as principal painters in the latter, now that Girolamo was dead, Muziano, the young Federico Zuccari and Agresti.[3]

In Ercole's time, particularly toward the end of his reign, the spirit of the incipient Counter-Reformation began to make itself felt in Ferrara.[4] Times had changed since mathematical and astronomical lectures at the university

---

[1] See, on these cartoons and on Giulio's antique models, Dollmayr, *op. cit.*

[2] Begarelli, whose art can in some ways be compared with that of Girolamo, was equally active for the Duke. As for sculptors from outside the Duchy, Cellini, who for a while was in Cardinal Ippolito's service, not only made the latter's bust and seal but also a medal of Ercole II (1540). The Duke also tried to bring Jacopo Sansovino into his service. He further ordered for Modena (after an apparent failure with Begarelli), about 1550, at a time when the artist had already turned toward mannerism, the giant statue of a Hercules, another example of his own apotheosis. Since Sansovino delayed the delivery of the statue, in 1552 the Duke commissioned a model from the artist's former pupil, Vittoria, who was penetrating Venetian sculpture with a consistent mannerism. Sansovino's statue, finally erected in Brescello, is almost Bandinelli-like, and was later taken for an antique statue.

[3] On Ippolito's activities as patron of art, see V. Pacifici, *Ippolito II d'Este*, Tivoli, 1920.

[4] There is perhaps no other Italian town in which the clash of Reformation and Counter-Reformation was outwardly so dramatic. It is well known that the Duchess, Renée of France, was herself a Calvinist and that Protestants from Italy and abroad, Marot, for

---

this late Venetian style of his, expressed not only in altarpieces, but in mythological paintings, in and outside palaces, which took the Duke's fancy. Perhaps it is worth re-calling here that Pordenone, in an effort to be also socially on a par with Titian, obtained the title of 'Cavaliere' from, *faute de mieux*, the insignificant King of Hungary.

were renowned enough to draw Copernicus there to study and (under Alfonso I) Pomponazzi could teach against the immortality of the soul. Now, erudite, but not too searching, humanists were to the fore[1] (though even they had occasionally to seek shelter behind Ercole from the reaction),[2] and neo-Aristotelianism, a literary movement in many respects connected with the Counter-Reformation, held sway among outstanding university teachers. Yet the influence of the Counter-Reformation on court culture, though constantly present in the atmosphere, was not yet all-pervasive under Ercole.[3] Nor does Girolamo da Carpi, who was very much a part of this court culture and whose art at this time was almost entirely of a secular character, with an unrestricted use of nudes, seem to have been deeply affected by it.[4]

[1] On the uncritical methods of the Ferrarese mythographers see J. Seznec, *La survivance des dieux antiques*, London, 1940, pp. 201 ff.

[2] Calcagnini himself, Cardinal Ippolito's tutor, who died as late as 1541 and was still alive at the beginning of Girolamo's activity for the court, stood between the two periods. He had been a friend of Erasmus and was one of the first to accept Copernicus' system. Later in life he became a priest and persuaded a Dominican to write against Luther's doctrines.

[3] See Appendix.

[4] It is interesting that, by mere chance, Girolamo was the only Italian artist working for the heretic Duchess. In 1557 (sometime after her release from confinement) she wrote to the Duke, asking him to authorize Girolamo to finish the paintings in the Castle of Consandolo which he had begun for her. (This letter furnishes proof that Girolamo was still alive in 1557 and could not have died in 1556, the year given by Vasari.)

instance, found refuge under her protection. Even Calvin visited her in Ferrara. After this Protestant wave of the thirties the reaction followed. In 1547 the Jesuits came to Ferrara and in 1550, through the initiative of Francesco Borgia, the College of Jesuits was erected and it was there that the Duke had his eldest son, the future Alfonso II, educated. Renée was for a short while confined by her husband to close quarters, and later forced by Alfonso to go back to France. Nevertheless, Ercole II, though a sincere Catholic and anxious to be on good terms with the Pope, on whom he greatly depended, was not a typical exponent of the Counter-Reformation. Still less so was Girolamo's special patron, the very easy-going and worldly-minded Cardinal Ippolito, who stood in open opposition to the zealous extremists within the College of Cardinals and who, when Papal Legate in France, did not refrain from listening to a Protestant sermon. Prohibited works in his library (e.g. Erasmus, Machiavelli, and many scientists) were burned by order of Cardinal Ghislieri when head of the Inquisition, and after he became Pope as Pius V, Ippolito fell openly into disgrace. He apparently came under the influence of the Jesuits only in his very last years. On the details of his career, see Pacifici, *op. cit.*

## Observations on Girolamo da Carpi

The representative of a very intellectual art, Girolamo was an intellectual himself, and quite naturally in close touch with the outstanding humanists of the university and the court, whom he painted and who praised him.[1] Most of the reputed professors of Ferrara University at the time of Ercole II were students of antiquity and were often employed as ambassadors and secretaries by the Duke. And it is significant that two of the three great Italian mythographers of the century lived close to the Ferrarese court and published their main works precisely during the years of Girolamo's activity there: Giglio Gregorio Giraldi (once a fellow pupil and still an old friend of Calcagnini) and Vincenzo Cartari.[2] These humanists considered as one of their main tasks the furnishing of themes and models for the artists, and we are, in fact, informed of the association of Girolamo with Giraldi, the most erudite and respected among them.[3] We shall later discuss in more detail Girolamo's close relation to antique sculpture—most of his drawings, for instance, are copies after ancient statuary—but it is obvious at first glance that the artist of this strongly humanist court, who painted all these smooth, statuesque mythological figures, approached the antique with an almost eru-

[1] Girolamo is said to have painted the following portraits: Antonio Tebaldeo, the famous poet; Francesco Alunno, a philologist; and Girolamo Falletti, the Duke's orator (the *Portrait of a Man*, Rome, Gallery Corsini, probably by Girolamo, is thought to represent Falletti). Falletti praised Girolamo in a poem, comparing him with Apelles. Concerning the courtly flattery in poems of Falletti and the famous humanist, Lilio Gregorio Giraldi, which refer to Girolamo's pictures and portraits of Anna d'Este, see note 1 p. 133 and various references in Serafini, *Girolamo da Carpi*.

[2] To complete the picture of the milieu in which Girolamo lived, we must mention Andrea Alciati, celebrated as a jurist and more important for us, as the creator of emblematic poetry inspired by antiquity and as the most influential author of the moralizing allegory of the *Emblemata*. He came to Ferrara on the Duke's pressing invitation and lectured at the university between 1542 and 1546; he even had his portrait painted by Battista Dossi at the special order of the Duke—a rare distinction—and he was offered the cardinal's hat by Paul III during the latter's visit to Ferrara.

[3] See, on these connections, the scattered references in Serafini, *op. cit.*, who found, for instance, that while Girolamo was in Rome he drew an Adonis sarcophagus (now in the Louvre), previously described by Giraldi. Though on the whole the mythographers seem scarcely to have taken notice of the concrete monuments themselves, apparently Giraldi paid somewhat greater attention to them. Wittkower (*op. cit.*) has demonstrated that the thematic details in Girolamo's *Chance and Penitence* originate in Giraldi's *De Deis Gentium*.

dite appreciation. When judging his mannerist but concentrated classicism, we must also take into account the formal ease (in contrast to the exhausting efforts of Garofalo and Battista Dossi) with which Girolamo expounds, in his allegorical and mythological pictures, the personal political characteristics of the Duke, following in detail, as was his custom, the scholarly precepts of the humanists.[1]

In everything of importance that was produced in the intellectual life of Ferrara at that time, Girolamo took part in a way that tends to throw some light on his art. I should therefore like to emphasize his well-known association with Giovanni Battista Giraldi, a younger relative of the above mentioned Giraldi, the outstanding literary personality in Ferrara between the time of Ariosto and Tasso. Giovanni Battista Giraldi was a poet, playwright, critic, professor of rhetoric at the university (he was also one of the Duke's secretaries), whose neo-Aristotelian reform of the drama, based on rules, played a significant part in Italian and in international intellectual life as well.[2] Girolamo constructed and painted the scenery for the famous performances, attended by the Duke, of Giraldi's two outstanding plays.[3] Since the presentation of Ariosto's comedies, these performances were the most important theatrical events in Ferrara, a milieu so reputed for its passionate love of the stage—all the more important since in no other large town in Italy were the dramas of contemporary playwrights performed. For Giraldi's tragedy, *Orbecche* (1541), neo-Aristotelian in theory but neo-Senecan in practice, the décor seems to have been a town with its palaces; for his *Egle* (1543), perhaps the very first pastoral play to be written,[4] the scenery was of a rural character, suitable for satyrs and nymphs. The conception of these plays, with their respect for antiquity and quest for new ways, elevated, at times genuinely human, somewhat sensual yet didactic and moralizing, is not only character-

[1] In the care with which he included the precepts of the humanists in his pictures, Girolamo resembled Titian and Giulio Romano, however different the three artists were stylistically.

[2] There is much good information on Giraldi's ideas in C. Guerrieri-Crocetti, *G. B. Giraldi ed il pensiero critico del sec. XVI*, Milan, 1932.

[3] For various other associations between Giovanni Battista Giraldi and Girolamo, see again the references in Gruyer and Serafini.

[4] *Egle* was the first in a long series of famous pastoral plays performed at the Farrarese court which culminated in Tasso's *Aminta* and Guarini's *Pastor Fido*.

istic of the notions current in Girolamo's circle but also offers a parallel to some aspects of his art.

Just as illustrative of Girolamo's art is his collaboration with Giovanni Battista Canani, another light of Ferrara university, the most celebrated Italian anatomist of his day.[1] Girolamo made scholarly drawings of arm and hand muscles for Canani's book, *Musculorum humani Corporis picturata Dissectio* (1541). The usual counterpart to the abstractness of a mannerist artist was the interest in rendering details most exactly, in this case, even in a scientific way—a fact which makes the facsimile-like exactitude of the mythological pictures and the Hampton Court portrait, with its evidence of great knowledge of facial structure, all the more understandable.

Ferrara with its court and university is the ever-present background of Girolamo's art. Even the milieu in which he lived in Rome between 1549 and 1554 was, in the wider sense, but an extension and intensification of the Ferrarese humanistic atmosphere. Although we shall treat the Roman episode separately, in order to draw attention to some of its particulars, it does not in itself seem to have caused a decisive break in his development but rather to have accentuated tendencies already existing. We are therefore justified in dealing with his works carried out in the service of the court of Ferrara as a group, particularly his few extant mythological and allegorical canvases made for the Duke,[2] whether we believe them to belong before or after his Roman stay. The present stage of research does not permit an accurate chronology of Girolamo's whole *œuvre*. It can only very tentatively be suggested that of the four pictures of its type in Dresden which are entirely by his own hand, the *Judith* seems to be the earliest, followed by the *Venus on the Eridanus*, and then the *Ganymede*, while the *Chance and Penitence* appears to be the latest. There is much near-classicism in all of them, yet not even such a concentrated composition as *Chance and Penitence*, one of the allegorical pictures illustrating the Duke's personal policy, can really be called classicist. A classicism which was consistently rationalistic and realistic could not evolve

---

[1] Canani, who later became a priest, was also a collector of antique statues.
[2] Girolamo also made mythological pictures for Cardinal Ippolito, who was intimately connected with French politics and was the permanent candidate of the French party for the Papacy; he sent a *Venus* (no longer extant) by Girolamo as a present to the King of France.

as the art of the Ferrarese court and of court humanism in the mid-sixteenth century. Girolamo's works are almost equally close to, and different from, both realistic and painterly classicism. The former had previously occurred only in Central Italy, temporarily under such favourable social, political and cultural conditions as prevailed, e.g., at Rome when Julius II was pursuing a secular national policy. Painterly classicism appeared in Venice, the only Italian state which in the sixteenth century could retain, to some extent, its position as a maritime and trading power.[1] In some ways, Girolamo's paintings form an interesting parallel to a certain classicist nuance in Florentine mannerism.[2] A decidedly mannerist flavour permeates them all. Their over-smooth realism, the somewhat sculptural treatment, the clear, cool colours, slightly discordant, the abstractly wrinkled-up folds of drapery are not the means for creating a genuinely classicist whole.

[1] Venetian art and that depending on Venice, though belonging fundamentally to painterly classicism or the early baroque, is frequently tinged with mannerism, and the numerous possibilities of this combination give this whole painting its peculiar charm and variety. But it is, of course, scarcely ever so outspokenly, one-sidedly manneristic, irrational, un-realistic as, for instance, Florentine painting. For the relation of Venetian and Florentine mannerism, in detail, see pp. 70.

[2] This explains why Girolamo's works have so often been attributed to Florentines, and vice versa. Not only has he been confused with Pontormo and Bronzino but also with Salviati (figure compositions) and even with a seventeenth-century Florentine like Van-nini. Since a graphic classicism and its various transformations play such a conspicuous part in Florentine painting, numerous, though never quite close, stylistic parallels to Girolamo are to be found in Florence. Some of his early works, for instance, bear a resemblance to Pierfrancesco di Jacopo, his contemporary (and a much weaker artist) who transformed Sarto's and Sogliani's classicism into a light, fallow-coloured, 'Hodler-like' plasticity. He is on the whole more abstract than Girolamo, but with similar partly fantastic, partly naturalistic cloud effects. Even such late works as the altarpieces of Santi di Tito, an artist almost forty years younger, seem to constitute a certain Florentine parallel, in the modelling (even in the types), to Girolamo's *Pentecost* (see also note 1, p. 145), Santi having 're-naturalized' the stonelike plasticity of his master, Bronzino; but Santi's late classicism is already slightly inclined toward baroque and consequently his pictures are more spatial than Girolamo's. In portraiture Pierfrancesco and Santi di Tito are far more insipid, the former, in particular, more flat and schematic than Girolamo. This also refers to portraits by Pierfrancesco which, even quite recently, have been incorrectly attributed to Pontormo (such as that of *Bartolommeo Compagni* in the Sterling Collection, Keir) by E. Toesca (*Il Pontormo*, Rome, 1943, fig. 54.).

## Observations on Girolamo da Carpi

At first, in Girolamo's Ferrarese works, there is still much similarity, of a general stylistic nature, not only to Giulio Romano but even to Parmigianino, though points of likeness to the latter are fewer than they were in Bologna. Still, the Parmigianinesque type of expression, so marked in Girolamo's Bolognese pictures, did not at once disappear in Ferrara. The *Judith*, also resembling the Hampton Court portrait, is more or less a variation of the *Madonna della Rosa* with her tender, dreamy sentiment. Moreover, though Girolamo's relation to Venetian painting is never very close, it is now relatively the nearest—in these years we are still in Dosso Dossi's Ferrara.[1] It is perhaps no exaggeration to perceive in Girolamo's *Judith* a distant reflection, however monumentalized, of Giorgione's *Judith* in the Hermitage.[2] Briefly, one might say that Girolamo brings out the near-classicist and moderately mannerist elements of Giulio Romano (whose mythological panels may have appealed to him more than the frescoes) and of Parmigianino, leaving aside their extreme mannerist features.[3] Girolamo's gracious undulating line, as we have already remarked, is more temperate than Parmigianino's linear rhythm and he never applies Giulio Romano's rather stormy ornamentality to the body but only to the draperies. The more these earlier influences wear thin or are modified, the more he works out, as in the *Ganymede* and in *Chance and Penitence*, his own ripe, concentrated style. Personal as this style is, it can be said to stand somewhere between Raphael and Peruzzi, but of course one generation later when it had felt the impact of mannerism.

Girolamo's historical position is already apparent in the *Venus on the*

---

[1] However, during these same years, the late Dosso himself was much more violent in expression than Girolamo. In his very late, grotesque *Stregoneria* (Pitti)—one of the most interesting documents of Ferrarese mannerism and a parallel to the contemporary mannerist-expressionist art of Germany—the smiling type of woman seen *en face* might be considered, in theory, as it were, a transformation of Girolamo's *Judith*.

[2] While discussing the Giorgionesque expression of Girolamo's *Judith*, it is perhaps worth noting that in the chapel of the Ducal Castle of Ferrara there was a *Judith* attributed to Leonardo (A. Venturi, 'Quadri in una cappella estense nel 1586', *Archivio storico dell'arte*, 1888, I, p. 426), which Girolamo must have known.

[3] Girolamo's altarpiece of *St. Jerome in the Desert*, in S. Paolo at Ferrara, also probably one of the earlier works, has its genesis in the same saint in Parmigianino's *Madonna of S. Margherita*, now turned into a full-length, slightly classicist figure.

*Eridanus*, which may have been painted in the early forties.[1] He has possibly used, in his cold but by no means unsensuous painting, motifs from Correggio's *Io* and *Leda*. Yet on the whole his picture is nearer to Raphael's and Peruzzi's frescoes in the Farnesina. The poise of Raphael's *Galatea* and of his *Venus and Psyche* cycle was probably, fundamentally, Girolamo's ideal. Peruzzi's frieze in the Sala delle Colonne of the Farnesina, however, with scenes (mostly) from the *Metamorphoses*,[2] among which the *Birth of Venus*[3] seems to be the mannerist-archaeological answer to the *Galatea*, is not far away as an historical precursor and parallel.[4] For Peruzzi, just as Girolamo, stands between classicism and mannerism, though nearer to the latter than Girolamo. Peruzzi does not build mere impressions of antique statues organically into a different, modern, classicist baroque world, as does Raphael, nor does he create, with masterly freedom *vis-à-vis* antique sculptures, sumptuous scenes of an early baroque character like Titian. Rather, in his mythological paintings he keeps near to the ancient statues and reliefs themselves,[5] preserving, in the spirit of early mannerism, many of their thematic and formal features. Even now Girolamo is too close to Raphael to go as far as Peruzzi. Yet an increasing nearness to Peruzzi and to Peruzzi's attitude toward the antique is discernible, particularly after Girolamo had been in Rome. This can be seen in the relation of Girolamo's *Ganymede* (Plate 37b), possibly

[1] Quite apart from stylistic considerations and for external reasons alone, Girolamo's picture must have been painted in the forties and probably not very late in that decade. For, in a poem published in 1546, Falletti, the humanist, compares this *Venus on the Eridanus* with Anna d'Este, the Duke's daughter, whose beauty was much praised at court and who married the Duke of Guise in 1548.

[2] These frescoes, representing a complete mythological encyclopædia, one of the decisive works in the development of mannerism, have been incorrectly attributed to Giulio Romano by F. Hermanin (*La Farnesina*, Bergamo, 1927, p. 83). Although Peruzzi's authorship was established long ago by Crowe and Cavalcaselle (*History of Painting in Italy*, London, 1866, III, p. 392) and by Dollmayr (*op. cit.*, p. 215), many scholars (among them myself) have thoughtlessly followed the more recent, incorrect attribution.

[3] Peruzzi's Venus in this scene is an anticipation of Beccafumi's ripe style.

[4] The extent of Girolamo's actual knowledge, before going to Rome, of Peruzzi's numerous compositions in the Farnesina, through various kinds of copies, is an open question.

[5] Concerning Peruzzi's antique models in the Farnesina, see R. Forster, *Farnesina Studien*, Rostock, 1880, and F. Saxl, *La fede astrologica di Agostino Chigi*, Rome, 1934.

painted after the Roman stay,[1] to Peruzzi's fresco of the same subject in the Sala di Galatea of the Farnesina (Plate 37a). Indeed, it is difficult to suppose that Girolamo did not know Peruzzi's work at the time of painting his own picture, though he does not follow it closely. The eagle's widespread wings, for instance, are more reminiscent of Michelangelo's than of Peruzzi's composition. Yet neither Michelangelo's mannerist violence nor the breath-taking baroque of Correggio's famous representations of this theme have really appealed to him here. The softly flowing line of Girolamo's *Ganymede* is apt to remind one of antique compositions of the same theme.[2] And in sixteenth-century art he was drawn to the easier flexibility and stronger classicist nuance of Peruzzi's composition, where the grace of the antique model is brought out so palpably in the delicate modelling of the body. By gently turning around Peruzzi's Ganymede, softening the contours, giving more compactness to the ornamentally fluttering draperies, even smoothing the waving of the hair in the wind, Girolamo created a composition which seems still more classicist, particularly more Raphaelesque.

This strong classicist note is just as evident in the *Chance and Penitence* (Plate 38a). The figure of Chance bears quite a strong resemblance to the central figure on a sheet of pen studies in the British Museum (Plate 39a), drawn by Girolamo in Rome, which is a copy after the Hellenistic statue of a Terpsichore[3] then in the Villa Giulia.[4] Although Girolamo could not have

[1] True, the shape and size of the *Ganymede* (though, perhaps not its theme) might suggest that it was intended for the same room as the two companion pictures of Battista Dossi's *Dream* and Battista's and Girolamo's *Hora*, the first certainly, the second, probably, finished before Girolamo went to Rome (see note 3, p. 135).

[2] In a very general way only is Girolamo's composition reminiscent of the statue, now in the Museo di S. Marco at Venice. In his picture the figure of Ganymede is more completely embedded in that of the bird, with a result somewhat comparable, for instance, to the relief on the famous Greek mirror in Berlin.

[3] On this, as Anchirrhoe, see both Reinach, *Répertoire de la statuaire*, I, p. 436 and B. Ashmole, *Catalogue of the Ancient Marbles at Ince Blundell Hall*, Oxford, 1929, no. 37. I am most grateful to Prof. B. Ashmole, Keeper, and to Mr. C. M. Robertson, Deputy-Keeper, of the Department of Greek and Roman Antiquities in the British Museum, who have kindly helped me to identify some of the antique statues mentioned in this article and have given me very useful advice on various questions relating to archaeology.

[4] Girolamo's note underneath the figure mentions the Pope's Vigna. However, it probably does not refer to Julius III's Vigna in the Via Flaminia but to the Villa itself (often also called Vigna), where the Pope had some thirty antique statues.

been entirely oblivious of Dürer's *Fortune* when he painted the figure of Chance,[1] who stands on a globe in mid-air above a landscape, one can assume, I think, with some probability, that he also used as a model the same antique statue copied in the drawing.[2] The slim figure of Terpsichore has become in the painting an equally slim male, whose attitude, though more composed, is similar to the statue, particularly one foot and the hand holding the drapery. While large batches of manneristic folds have been added on either side, the drapery around the body has been simplified but the exposed right breast retained by means of the diagonally folded material.[3] On the whole, the

[1] Wittkower (*op. cit.*) has already noticed this connection.

[2] If one regards the drawing, certainly made in Rome, as the concrete model for the painting, another proof would, of course, be added for the late dating of the latter. For Wittkower (*op. cit.*), who has also given good reasons for interpreting the picture as 'Chance and Penitence' and not, as traditionally, as 'Chance and Patience', concluded on the basis of iconographic evidence that the painting dates from the fifties. According to him, the Dresden picture cannot be the *Chance and Patience* to which an account of 1541 refers, so that the earlier painting is apparently no longer extant. Nevertheless, I think Wittkower goes too far in suggesting, for the sake of his theory, that all stylistic questions still remain unsettled in Ferrarese painting of this period and that anything can be assumed. Thus he leaves open the attribution of the two pictures in Dresden, representing *Peace and Justice*, which he surmises (following A. Venturi) must have belonged, together with Girolamo's late *Chance and Penitence*, to the decoration of the same room. However that may be, one cannot seriously doubt that these two paintings are by Battista Dosso (though Girolamo's collaboration, as on many of Battista's pictures, is certainly possible); since he died as early as 1548, it is quite admissible to assume, as Mendelsohn does (*op. cit.*, p. 164), that an account of 1544 refers to the *Justice* in Dresden.

[3] Girolamo's treatment of draperies is very varied. There is usually much of Peruzzi and Giulio Romano behind it but sometimes he transforms these original influences in a more naturalistic, sometimes in a more anti-naturalistic way, and usually in a combination of both. The last seems to be the case in a most striking mythological picture in Dresden which, though one of the few Ferrarese paintings in the Gallery not attributed to Girolamo by Serafini, is, I believe, in great part, by our artist. It represents *One of the Horae with the Steeds of Apollo* (Plate 36b), the companionpiece (*Awakening*) to Battista Dossi's late *Dream*, and is far too good and graceful a picture to merit the very indifferent attribution, to the 'School of Battista Dosso', at present accorded it. While the composition may be by Battista, the type is rather that of Girolamo's drawings after the antique, and the whole execution, with its economy of detail, bears strong imprints of Girolamo. Typical of him is the elegant and agitated drapery which retains Peruzzi's ornamentality and Giulio Romano's puffed-out effects, and is given an even more genuinely 'wind-blown'

picture, with its concentrated composition, is more serene *vis-à-vis* the some-what agitated antique model and Girolamo's own, very spirited drawing after it. Here, as always, he gives the impression (confirmed by our knowledge of his antique models) of having made the very stone of the statues come to life. This quality appears, in Girolamo's customary manner, in the naturalistic, lucid, precise (yet not really Bronzinesque) modelling, in the individualizing of the types (though not quite so marked as in his early days), in the quiet, demure sentiment. Even the sad, calmly resigned expression of Penitence can be felt as a more generalized counterpart, in the world of allegories, to the expression of the Hampton Court portrait.

To round off our survey of Girolamo's more interesting works done for the court, I must further refer to three frescoes with Bacchic themes, still extant in the Ducal Castle. Their present, much restored, condition makes it difficult to determine precisely their date (and the degree of assistance Girolamo received from others) but their combination of Titianesque and particularly Giulio Romano-like features places them perhaps among Girolamo's

---

appearance. (Historically speaking, this drapery style, in Girolamo and in Peruzzi—who, rather than Giulio, probably painted the *Apollo and the Muses* in the Pitti—derives from the late Mantegna's dancing Muses in the *Parnassus*.) Typical of Girolamo also are the enormous, very real clouds, and the cold, light colouring of the celestial atmosphere with its trenchant grey-blue. One can also compare the use of the olive tree in the Dresden picture with a work which is entirely by Battista, such as the *Minerva and Cupid* in the Donaldson Collection: in the latter, it is displayed in a rather thin manner; in the former, it is much more organic and succulent, accentuating the place of the figure in the composition. Various drawings of Girolamo after Amazons and Maenads from Roman sarcophagi come surprisingly near to the figure of the Hora in type, attitude, and draperies, though the draperies in these more or less literal copies have not quite that naturalistic motivation which Girolamo carries through in the picture. Particularly close to the Hora are two pen drawings by Girolamo, one in the Albertina (Catalogue *cit.*, no. 89, incorrectly attributed to Giulio Romano), and the other in the British Museum, both copying the same fighting Amazon of a third-century sarcophagus which represents the *Battle between Achilles and Penthesilea* (Reinach, *Répertoire des reliefs*, III, p. 352) and was, in Girolamo's time, in the Villa Giulia (now in the Vatican). In the Vienna drawing, Girolamo has copied not only this favourite Amazon figure but a part of the fighting scene. He has even added more heads of horses than in the sarcophagus itself, so that the similarity with the Dresden picture is very great indeed. All the same, though Girolamo must have made these drawings in Rome, it is probable that he finished the picture before going there in 1549, perhaps after Battista Dossi's death in 1548.

relatively early Ferrarese works. They could derive approximately from the same years in which Titian himself underwent a certain influence from Giulio Romano's frescoes in Mantua.[1] Thus Girolamo's frescoes seem to reflect this contemporary, equally North Italian, stylistic combination. The frescoes, more pretentious than any of his works discussed thus far—their scholarly conception doubtlessly going back to the court humanists—are rather different in quality. Of the three, it is the charming idyllic *Vintage* which (though a strong flavour of Giulio Romano is evident) seems to be the most characteristic of his personal style and which demonstrates with what grace and sensitiveness he can now build lively movements into the pattern of even a large scene when the theme demands it. The *Triumph of Ariadne* is the least characteristic of his style. Here Girolamo has naturally been influenced by motifs in Titian's picture of a similar subject, then in the Castle (now in London), but was even more drawn towards Giulio Romano. The embracing couple in the foreground, particularly the woman, is, I believe, derived from Giulio Romano's group of *Bacchus and Ariadne* in the Sala di Psiche of the Palazzo del Tè. The third fresco, the *Triumph of Bacchus* (Plate 41b), is only a free copy after Garofalo's previously mentioned picture of the same theme (Plate 40), also in the Castle. Girolamo was certainly ordered to keep close to this painting, so famous at the time. But since this free copy gave him an opportunity to criticize, within limits, the composition of his former master, twenty years his senior, it affords us a good insight into the artistic aspirations of the more sensitive Girolamo, who belonged to a new generation. For this reason and because we can reconstruct the exact story of its origin, we discuss this particular fresco in some detail, although within Girolamo's *œuvre* it is a work of only secondary importance. To Girolamo, Garofalo's composition must have seemed far too congested, an enormous polyp-like group pressed together, inorganically, in one layer, and he tried all means to clarify and to organize it. This does not mean that Girolamo was a *pur-sang* classicist; but it does mean that he was not satisfied with the mere 'embarrassment solution' of the elderly Garofalo, who was fundamentally a classicist of an early, old-fashioned, primitive type and a mannerist only *malgré lui*. The process of clarification and concentration is seen, more intimately than in the

---

[1] See Hetzer, *op. cit.*

ruined fresco, in a pen drawing (Plate 41a) which I have found in the British Museum. It is a study by Girolamo after Garofalo's picture and so represents a stage previous to the fresco, where one can follow closely how Girolamo first felt his way about Garofalo's pattern, trying alterations here and there. He attempted to bring out some main groups and to subordinate the individual figures to them. For instance, the young boy in Garofalo's picture, squeezed behind the elephant's trunk, becomes much taller and the panther is pushed slightly to the side, so that the boy's long, clear contours can play a larger part within the pattern of the whole. This figure rounds off the group to the left and also forms a real transition to the Silene group to the right. The man steering an elephant is raised higher than the three mounted boys to the right of him, in order to counterbalance them. Girolamo has lifted the embracing couple out of the throng and put it on the empty left side of the sheet so that he can study it freely. To loosen up the composition, he has also attempted to create space. Starting from the left front corner, he has pushed all consecutive figures increasingly inward so that the Silene group is left with a large expanse in front of it. Where he has not given real space, he has put the figures into strong relief, in the customary mannerist way, by making them stand out from a dark, closely hatched background, as, for instance, the woman holding the basket of fruit on her head. The latter has also been transformed into more of an antique than Garofalo's figure, namely, into a priestess of Bacchus, particularly frequent on Roman sarcophagi, or of Ceres. So Garofalo's frozen, porcelain-like figures have become both more naturalistically alive and more sculpturesque in the antique manner. In the fresco itself, where Girolamo has changed Garofalo's horizontal shape to a vertical one and has practically produced a work of his own, not only has he kept most of the detailed alterations of the drawing but his procedure of concentration and loosening has been much more radical. He has cut off the whole right side of Garofalo's overcrowded gathering (perhaps following, in this, the original Raphael composition which had served as model for Garofalo), reduced the number of figures to the more significant, made a distinction, more pronounced than Garofalo, between the two layers of the foreground and background figures, replaced Garofalo's usual zig-zagging mountain landscape with a more even and harmonious one. Above all, through the addition of a very wide foreground and a high sky, he has entirely changed the relation of the figures to the foreground and the sky. Yet precisely these large spaces above and below form not only a counterpoise (for the sake of which the group of Jupiter and Juno in the clouds has been raised much higher, as has the flying genius on the left) but an intentional contrast to the still relatively compressed strip of figures, so that the over-all effect is not in the least three-dimensional. Consequently, while, on the one hand, Girolamo has created a more organic composition, whose traits of ripe classicism were beyond the reach of Garofalo, on the other hand, his usual mannerist features are not lost to view.

To discover Girolamo's relation, in his ripe age, to contemporary art, more alive and modern than Garofalo's, we must turn to his stay in Rome, where he had, though tardily, opportunity for contact with it. There also, after his early, indirect experience of the antique through Peruzzi's Bolognese cartoon and Giulio Romano's frescoes in Mantua, after his constant collaboration with the humanists of the court of Ferrara,[1] and with his limited knowledge of antique originals, he finally came, at the age of forty-seven, into direct touch with most of the famous ancient monuments themselves. This episode, of which at present we are little informed, is so interesting a period in Girolamo's life that a more detailed excursion into it may help us to understand his art, particularly in its late phase.

Girolamo's patron, Cardinal Ippolito, was the outstanding collector of antiques in Rome,[2] and Girolamo, as an expert, assisted him on his famous excavations of antique statues in Rome and Tivoli.[3] In Ippolito's gardens on

[1] In the case of Girolamo's decoration of the ceiling of the Camera dell'Aurora (1548) in the Ducal Castle, representing, in elaborate mythological terms, the *Story of the Day*, the particulars were also, of course, given by the humanists. Though the paintings were largely executed by the much clumsier Camillo Filippi, they reveal, even more than the frescoes with Bacchic themes, Girolamo's very obvious studies after antique statues. The movements and the draperies of the figures are slightly under the influence of Giulio Romano and, in an historical sense, they resemble even those of Piombo in the Farnesina.
[2] A taste which his brother, the Duke, shared with him, at least as far as Hercules statues were concerned.
[3] Concerning Ippolito's large collection of antiques, see P. G. Hübner, *Le statue di Roma*, Leipzig, 1912, pp. 108 ff. and C. Hülsen, *Römische Antikengärten des XVI. Jahrhunderts*, Heidelberg, 1912, pp. 90 ff.

the Monte Quirinale Girolamo built small wooden temples, covered with foliage, to contain antique statues—a novel and astonishing spectacle in Rome.[1] As an architect, Girolamo was also influenced by Peruzzi[2] and, in architecture as well, combined classicist and mannerist features in his style. And as an architect, he worked in Rome not only for Cardinal Ippolito but, on his recommendation, for Pope Julius III in the Belvedere, precisely where the most famous collection of antique sculptures then extant was located. So Girolamo found himself suddenly in the very centre of Roman artistic life. He was in the service of the prominent patrons of architecture and lived, moreover, as their employee, amid the greatest possible wealth of ancient statuary (the Villa Giulia, one of the most important secular buildings of the century, was put up by the Pope to house his private collection of antique statues). In fact, in Rome Girolamo assiduously made innumerable drawings after antique works, perhaps more than almost any other important Italian painter of his time. He even copied the same sarcophagus or parts of it twice, so great was his interest.[3] For, like Peruzzi and Giulio Romano shortly before and Vasari and Pirro Ligorio soon after him, Girolamo represented a certain, though by no means common, type of artist of those decades: learned and many-sided, painter as well as architect, with an interest in, and

[1] Hülsen (*op. cit.*, p. 93) asserts that considerations of erudition appear to have played a smaller part for Cardinal Ippolito than for some of the other big collectors in Rome one or two decades previously, and that a relevant consideration for the Cardinal was the effective placing of the statues within the disposition of the garden. This is true, but, as appears from one of his letters, the Cardinal was greatly interested in the quality of his statues and relied, for this, on Girolamo's judgment.

[2] No record has come down to us of Girolamo's stage architecture, but I could imagine that here, too, he was under the influence of Peruzzi, an innovator in this field as well. As we know from Vasari, Jacopo Meleghini, a fellow-countryman of Girolamo, owned many Peruzzi drawings (others were inherited by Serlio), which Girolamo might have seen; for Meleghini lived in Rome, was custodian of the antiquities in the Vatican and one of the Pope's architects, shortly before Girolamo (he died sometime between 1549 and 1553).

[3] In the British Museum, for instance, are two of his drawings from a sarcophagus with the *Judgment of Paris*, in the Villa Medici: one of the whole, one of the right side only. Regarding his two copies of a sarcophagus with a *Battle between Achilles and Penthesilea*, see note 3, p. 135.

knowledge of, archaeology[1] which was reflected in painting in more or less mannerist features.

In Rome Girolamo copied, almost as if they were antique reliefs, Polidoro's façade paintings which themselves owed so much to Peruzzi's achievements (a drawing after that in the Via della Maschera d'Oro is in the British Museum). But most of Girolamo's Roman drawings are copies after actual sculptures. As one would expect, they frequently represent rather tranquil figures, where the formal solution seems to have interested him, but it is curious that he also manifests a certain predilection for figures in motion like Maenads and Amazons—the statue of a dancing Terpsichore (belonging to Julius III) has already been mentioned in connection with the *Chance and Penitence*. The drawings are for the most part literal copies, with slight variations here and there. The attitudes tend to become somewhat more restless, the figures more slender, the draperies richer, while the faces usually appear very alive and bold—indeed, more alive than the originals. Thus the drawings reveal again, and more intimately than the paintings, the way in which he made the 'stones come to life'. Girolamo is here more dashing than in his paintings, which were intended for general view, and where he restrained himself for the sake of the formal pattern of the whole. Moreover, these drawings also demonstrate an unusually wide sphere of interest within antique art. It is characteristic of his taste, for instance, that he chose to copy one of the refined archaistic sculptures, fashionable in Rome in the first centuries B.C. and A.D.,[2] although not a great many of them could have been extant

---

[1] Perhaps I may mention here that the well-known type of artist, domiciled in Rome, who made a profitable business for himself by acquiring antique statues for art lovers, also originated in such a milieu, with Giulio Romano. In 1520 he bought the Ciampolini Collection of antiques and sold various pieces of it to Pope Clement VII and to the court of Mantua.

[2] It is curious that at the time of Augustan classicism there was also great interest in Rome in neo-Attic reliefs as well as in archaistic art (on the chronology of the latter, see G. Becatti, 'Lo stile arcaistico', *Critica d'arte*, VI, 1941, pp. 32 ff.), perhaps—I only hesitantly propose—as a somewhat less classical corollary rather than as a contrast. Toward the middle of the sixteenth century the artistic situation in Italy, superficially speaking, was in some ways similar. In so far as any classicism existed, as, for instance, in Girolamo, it also had a mannerist flavour. And for mannerism (just as previously for Quattrocento gothic) it was, of course, the neo-Attic reliefs with their ornamental, rhythmically

in Rome at that time. His drawing (British Museum) (Plate 44a) shows a careful study of one of the richest of the archaistic statues, one of the Graces, a relief figure on a triangular base (now in the Louvre) and a fairly close imitation of the Greek style of the end of the sixth century.[1] The study of archaistic folds, with scallop-like edgings, just as of abstractly floating, coiled draperies, abundant in all phases of Roman Imperial art, doubtless helped him to make his own mannerist drapery more ornamental and capricious in a rather individual and interesting way. On the whole, Girolamo's drawings after the antique, usually pen drawings, sometimes rendered more impressive through an imaginative use of wash, have a very cultured and original character of their own.[2] Since one cannot separate these drawings from Girolamo's cultural background of the learned Ferrarese court humanism, it is perhaps not too far-fetched to sense in their occasional bizarre character some reflection of the rather fantastic penchant of this humanism for late antique and mediaeval text sources, and for oriental and pseudo-antique deities.[3] Thus it is natural to be reminded of G. B. Giraldi's neo-Senecan, terrifying but measured, tragedies (for which Girolamo painted scenery) when confronted by Girolamo's sophisticated, agitated, tense figures of the Maenad and Amazon type, derived with heightened expression from Roman reliefs. Nevertheless, in spite of their personal note, Girolamo's copies after

[1] See Reinach, *Répertoire de la statuaire*, I, p. 65. The Louvre acquired from the Borghese Collection the base, which has two tiers of decoration on each face; in Girolamo's time it must have been in Rome. Winckelmann took it for a genuinely archaic work and it gained some fame under the title of 'Altar of the Twelve Gods'.

[2] In Serafini (*op. cit.*) the antique originals of only a few of Girolamo's copies have been identified. When someone (more versed in archaeology and the history of archaeology than I) has examined, in detail, Girolamo's studies after antique monuments, so that we know exactly what models he chose, how far he transformed them, and what his reaction was to the various styles in ancient sculpture, we shall have, I believe, quite a helpful clue to Girolamo's development, particularly to his late style. Besides, Girolamo's numerous drawings (as also those of Franco) constitute an almost untapped source of wealth for helping to establish the inventory of antique statues extant in Rome about the middle of the century.

[3] See Seznec, *op. cit.*

waving draperies which were the given source of inspiration. Girolamo's copy which we are discussing shows that the more refined artists had a taste for archaistic art also.

the antique remain, by and large, Roman drawings of the middle of the sixteenth century. Not only have they much in common with those of Giulio Romano,[1] who was fundamentally the basis for all these artists' studies, but, because of a certain similarity in their technique and handling of line, they are even more often mistaken for those of Franco (who systematically copied the antique monuments in Rome, with the intention of gathering them in a book). Girolamo's drawings, however, are as a rule not only more original (even when they are copies) but also more plastic in character, more powerfully constructed, and one feels a deeper comprehension of nature behind them. Those of Franco (even when they are not copies) are more the work of a pseudo-genius (they sometimes look like anticipations of Fuseli),[2] more playful, superficial, two-dimensional, schematically ornamental. The frequent similarity of Girolamo's and Franco's drawings after the antique, done in Rome in the same years, makes the stylistic contacts between them evident. A sheet in the British Museum, with drawings on both sides, even suggests the probability of personal relations between them. The pen drawing on the back of the sheet, a copy after the antique statue of a river god, is undoubtedly by Girolamo's hand. The drawing on the other side (Plate 42a) is closely connected with the imposing figure of a seated, woman spectator on the left in Franco's fresco of the *Capture of the Baptist* in S. Giovanni Decollato (1538) (Plate 42b). Although the borderline between Girolamo and Franco is particularly narrow here, I am inclined to think the drawing is an original sketch by Franco, showing his facile, flowing line and rather timid wash. Thus Girolamo apparently owned a drawing by Franco,

[1] How similar Girolamo's drawings are to those of Giulio Romano, inspired by the antique, and yet how easy it is to distinguish between them, particularly in technique, is well demonstrated by the following. I have already mentioned a typical drawing by Girolamo (note 3, p. 135) which renders parts of an Amazon Battle. It is reproduced in the Albertina catalogue (*op. cit.*, no. 89) under the (wrong) attribution to Giulio Romano (although Dollmayr had already recognized it to be by Girolamo), while the drawings around it are works just as typical of Giulio. Far more different are Girolamo's drawings from those by Parmigianino. The Parmigianinesque drawings which Serafini attributes to Girolamo are mostly by Parmigianino himself.

[2] This is not just a coincidence. Early romantic artists, standing between classicism and mannerism (see, on this style, my observations on Girodet: 'Classicism and Romanticism' pp. 14 *seq.*) necessarily show similarities with their sixteenth-century ancestors.

made by the artist more than a decade before Girolamo's arrival in Rome—
he could have received it as a gift from Franco—and used the back of the
sheet for a small study of his own. If the alternative is true and the drawing
is a copy by Girolamo himself after Franco's fresco[1]—of which I am not
convinced—the relation between the two artists would, of course, be even
closer. But even so, Girolamo would scarcely have possessed this drawing
had he not a predilection for the closed statuesque motif of this particular
Franco figure, which is a combination of two or three of Michelangelo's
Sibyls and Ancestors. Since it gives us an idea of Girolamo's taste during
these years, with respect to contemporary mannerist art in Rome, the drawing is here reproduced. In it the Michelangelesque draperies are more closely
adhered to than in the fresco itself, where the garment has become richer
(this radical difference being a further argument in favour of Franco's authorship). Girolamo also copied Michelangelo's figures directly. That he combined in a drawing of a seated Prophet (Uffizi) at least two of Michelangelo's
Prophets is known. Moreover, I have come across, in the British Museum,
copies by Girolamo after two of the Ancestors: Asa and Roboam. In Franco's
derivation from the Sistine frescoes and in the originals themselves, the same
fundamentally simple problem interested him, for which he tried to find a
solution in his own lucid art: that of a more or less tranquil, monumental,
well-balanced, compact composition of one or two statuesque figures. That
is why the general effects of these copies of Girolamo and the Franco drawing
are so similar. Girolamo does not accentuate the cubist or even the fantastic
possibilities of the Ancestors as do the extreme mannerists of the Rosso type.
He is inclined to seek out, at least to a great extent, the potential classicism
of Michelangelo's composition. Therefore, although his copies are almost
literal, he straightens slightly the backs of his figures, thus easing the super-
human heaviness and architectonic pressure created by Michelangelo partly
by means of the triangular shape of the spandrels. Though Girolamo's keen
intellectual interest in statuesque motifs is mainly directed toward the enormous wealth of the antique, he was naturally not a stranger in Rome either
to the Sistine Chapel (on the other side of the sheet with the copy after a
Roman sarcophagus is the copy of the Ancestors) or to contemporary art.

[1] This is the British Museum attribution.

He apparently felt close, in some ways, to the mannerist artists of the Franco type, mainly Tuscan, who dominated Rome's artistic life in the years he was settled there. These men aimed at making their large mural decorations, painted in a Florentine-Michelangelesque style, more monumental, clever, and interesting by means of more or less obvious borrowings not only from the Sistine Chapel but from antique sculptures. It was at that time that he made friends with Vasari, who was also employed by Julius III and was one of the main exponents of this style[1] (although merely a weak echo of Salviati, the outstanding artistic personality of Rome in those years).

Perhaps most significant of Girolamo's not quite average taste is his having copied, in Rome, a curious fifteenth-century statue. On the same sheet as the copy after a Terpsichore statue (Plate 39a) is the drawing of a figure which stood on the Ponte S. Angelo, as Girolamo's own note beneath it indicates. It seems certain that this was the *St. Paul* by Paolo Romano (Plate 39b),[2] which stands today on the same spot. This sculpture, originally made in 1464, for Pius II's Loggia of Benediction, was rediscovered in St. Peter's, in a neglected state, by Clement VII, who, as Vasari says in relating the event, had good artistic judgment and put it up (together with a recently commissioned companionpiece of *St. Peter* by Lorenzetto) on the Ponte S. Angelo.

[1] I cannot help feeling that in his *Coronation of the Virgin*, in Città di Castello (1561, S. Francesco), Vasari may have been influenced by Girolamo's *Pentecost* in Rovigo, which he had apparently seen and which he praised in the *Vite* ('for the composition and the beauty of the heads'). While attitude, gesture, drapery of the seated Madonna (even the composition as a whole) are somewhat related, it is the type of the Madonna in Vasari which, 'Florentine' though it is, seems to me the nearest to Girolamo, originating with him rather than with Bronzino. It also appears to me that Girolamo's subtle drawings after the antique had a certain influence on various artists working in Rome, both Tuscans and non-Tuscans. I mention only Tibaldi, who was painting in the Belvedere during the years Girolamo was its architect. Consequently, it is probably not chance that some of Tibaldi's drawings which copy the antique or are inspired by it have much in common with those executed by Girolamo at the same time. That is why, for instance, a pen and wash drawing after a Flora statue, previously in the Poynter Collection, which, in my opinion, is certainly a work of Tibaldi, has been attributed to Girolamo.

[2] The resemblance of the drawing to the Paolo Romano statue is so strong that the only other alternative—namely, that Girolamo copied one of the apostle figures which Montelupo had erected on the same bridge in 1536, on the occasion of the entry of Charles V to Rome—is ruled out. I agree in this with Mr. Pouncey.

## Observations on Girolamo da Carpi

Art-historically, Clement VII's and Girolamo da Carpi's predilection for Paolo Romano is illustrative. For Paolo Romano, with the somewhat older Isaia da Pisa and a few others, belongs to that interesting group of sculptors, partly Tuscan, working in Rome as successors of Filarete, in the third quarter of the Quattrocento. They ostentatiously hark back to the drapery style of the late, occasionally even of the very late, antique[1] and their style could, to simplify, be designated as both classicist and pre-mannerist.[2] In Isaia da Pisa the stress is on the latter, while Paolo Romano is somewhat more of a classicist. Girolamo sensed that which was topical for him in this older art. It is characteristic that he should have drawn Paolo's statue side-view, thus obtaining a more closed motif (such as in the Franco figure as well). He even harmonized, by slight alterations, the attributes of book and sword, thus giving the whole an almost classicist effect.

Only the drawings offer a safe guide to Girolamo's Roman style,[3] but as most of those which have come down to us seem to have been done in Rome, they provide a good measure of his taste at that time. Among the drawings are a few which, though not copies after the antique, are near to it in spirit. Of these, a *Holy Family* in the British Museum (Plate 43) is, I believe, the most interesting and revealing.[4] One might almost say that in some of his Roman drawings Girolamo created a secret, personal world for himself, where certain bizarre features, which previously occurred only

[1] Sometimes, perhaps, they have also been influenced by mediaeval, Romanesque sculpture.

[2] How features of this style, later fused with that of Mino da Fiesole, working in Rome, crop up again in mannerist sculpture, particularly in that of Bandinelli (heads, draperies), has still to be worked out.

[3] We may, in time, become better informed than we are at present about Girolamo's paintings made in Rome, but even so it is very probable that quantitatively his output was rather inconsiderable. I am not quite sure on what grounds the ceiling decorations with mythological themes in the Palazzo Spada have been attributed to Girolamo by Serafini (*op. cit.*) and Gamba (article on Girolamo da Carpi in the *Enciclopedia Italiana*, XVII, 1933); but even given documentary evidence, Girolamo could only have had the direction of the work, as the rather weak paintings reveal different hands. Only very few of them show a certain resemblance with Girolamo while others are near to Stradanus.

[4] Even if this drawing was not made in Rome, it would seem to be the result of his Roman inspirations. The repetition of the composition by Ligorio, mentioned in the text, also indicates that it was executed during the Roman stay.

occasionally, are intensified,[1] even, as in this particular case, in a religious subject. The drawing seems to be a kind of religious and archaeological fantasy. The figures, though not direct copies, appear to be inspired by antique statues, and their stonelike treatment is intended to recall sculptures, as in so many other contemporary drawings made in Rome. The construction, the individual poses and gestures remind one of various extremely well-known Italian paintings. The drawing is perhaps closest to Correggio's *Marriage of St. Catherine* in the Louvre (which Girolamo probably copied in Modena and which had already influenced his S. Salvatore altarpiece) and to Parmigianino's *Holy Family* in the Uffizi. But these models have been even more transformed than the antique statues.[2] As frequently in mannerism, the Madonna is something of a cult-image, not mediaeval in character, however, as in Florentine mannerism but, as one would expect, like an antique goddess on an antique throne. Her face, not at all inhuman, is Girolamo's usual personal variation of the Augustan type. The St. Joseph, more than the corresponding figure in the Modena *Adoration*, has a Peruzzian antique head. At least so it appears. Actually, however, it is derived from a type of Parmigianino, in particular, from the old man, seen *en face* with fluttering hair and beard, in his etching of a *Boy Sitting beside Two Old Men*.[3] Girolamo's type is not a literal copy of this unkempt philosopher but, through small alterations, has acquired a more massive, blocklike, antique appearance, nobler, with a high forehead and less dishevelled hair. The suggestions for these changes may have come from innumerable sources. Apart from the usual ones of Raphael (*Ezekiel*) and Peruzzi, it is more than sufficient to mention the late Michelangelo (London cartoon of the *Holy Family*) and the antique itself (Nile statue). Most astonishing, however, is the figure to the left, which does not even pretend to be a holy figure but

[1] There is, for instance, in the British Museum, a drawing by Girolamo after the antique of the grotesque face of a satyr with a twisted lower lip.

[2] On the various possible combinations of archaeological knowledge with a free, arbitrary attitude toward antiquity, to be found in mannerist artists of antiquarian tendency, see, for instance, M. Dvořák, *Geschichte der italienischen Kunst*, Munich, 1928, II, pp. 122 ff. and E. Panofsky, *Herkules am Scheidewege*, Leipzig, 1930, p. 33.

[3] The drapery and even the attitude of the boy, in this etching, may also have influenced Girolamo's *St. Joseph*.

is an antique statue pure and simple, quite possibly a creation of Girolamo's fancy. She is an antique goddess with a Phrygian cap on her head, a basket in her hand, a kind of Abundantia or Cybele.[1] In general attitude she is perhaps influenced by the Minerva figures on sarcophagi of the second and third centuries,[2] and is conceived by Girolamo, more explicitly than before, in the style of the late antique, so congenial to mannerism. Another feature, significant for mannerism and for its particular interpretation of the antique, is the disproportionate scale of the figures, such as the small-sized foreground figure and the large St. Joseph in the background. But there are also obvious pointers to this female figure in mannerism itself. The Phrygian cap with its undulating shape is a frequent motif in Parmigianino's drawings and particularly in his etchings (and it was apparently from him that it passed to other artists who were more or less near to Girolamo in Rome, such as Salviati, Conte, and Franco).[3] There even exists an engraving by Vorsterman after a lost composition of Parmigianino of the *Presentation in the Temple* in which there is a woman, seated on the steps leading to the altar, with both a Phrygian cap on her head and a fruit basket in her hand. It is tempting to assume that this figure was the direct inspiration for Girolamo. The position of the figure recalls perhaps the Magdalen in Parmigianino's *Holy Family* in the Uffizi or the angel on the extreme left in the *Madonna del Collo Lungo*, both of which occupy a similar place in the composition, are in profile, and hold vases very ostentatiously in their hands. If there was any direct connection, this kind of pagan-archaeological sharpening and transformation of Parmigianino's figures and features is all the more interesting.[4] Girolamo would

[1] Prof. F. Saxl believed that this figure may have a more allegorical meaning.

[2] Professor Ashmole points particularly to a sarcophagus with the *Story of Proserpina* in the Villa Giustiniani.

[3] Girolamo himself already used the motif of the Phrygian cap before going to Rome, namely, on the ceiling of the Camera dell'Aurora.

[4] It amounts to the same thing that when viewed from the antique source, it appears to be a christianization. Mrs. H. Bober, New York, does not exclude the possibility that it is a christian adaptation of some lost antique painting, representing Cybele and Cronos. She rightly indicates that there are various motifs, among them the drapery of the Madonna and variations in scale of the figures, which point in this direction. Should the figure with the Phrygian cap have a prototype in antique sculpture, she believes it might eventually have been a Trojan woman from some scene of the Trojan War.

certainly never have dreamt of using such a composition in painting. Even so, it could have been produced in Rome, at the latest at the time of Julius III, called banally, but with some truth, the last Renaissance Pope and whose cultural milieu, in spite of the rapidly approaching Counter-Reformation, was still strongly infused with pagan-humanist elements, just as was that of the Ferrarese court.[1]

In a composition such as this *Holy Family*, Girolamo has drawn the consequences of Peruzzi's Bolognese cartoon more consistently than in his early period. In his late phase, as we have already noticed in discussing the *Ganymede*, Girolamo moved closer than ever to Peruzzi, whose spirit and the consequences of whose art were so vividly alive in contemporary art in Rome.[2] It is significant that a repetition of Girolamo's bizarre *Holy Family* occurs in a drawing at Chatsworth by Pirro Ligorio,[3] one of the most fan-

---

[1] This is even conceded by L. Pastor, *Geschichte der Päpste*, VI, Freiburg, 1913, pp. 48, 240, 257, and *passim*. It was this worldly spirit which led the Pope to commission Taddeo Zuccaro to paint frescoes of a bucolic character with fauns and nymphs in the Villa Giulia. Taddeo apparently went to Verona in 1551, just before he painted these frescoes, and on this North Italian journey very probably had occasion to see Girolamo's frescoes with Bacchic themes in Ferrara, to which his own frescoes show, in spirit, a certain resemblance.
[2] Vasari rightly says of Peruzzi (who was buried in the Pantheon, near Raphael) that his fame was greater after his death than during his lifetime. Perhaps I should mention that a pure, abstract plasticity is even more strikingly apparent in the drawings of Daniele da Volterra, who received his training under Peruzzi in Siena, than in the composition by Girolamo just discussed. It is assumed that Daniele came to Rome with Peruzzi in 1535, but the probability has not been considered that just before this date—as Mr. Pouncey and I became aware in reconstructing Peruzzi's development—Daniele may even have worked under Peruzzi (with, I believe, the old Pacchia and the still older Pacchiarotto) on the frescoes at Belcaro, painting the small mythological scenes. In view of their close personal connection, it is pretty evident that Daniele's first large work in Rome, the frieze with scenes from the *Life of Fabius Maximus* in the Palazzo Massimi delle Colonne, must have been procured for him by Peruzzi himself, the architect of the palace, shortly before his death in 1536. The frescoes reveal elements of the styles not only of Beccafumi and Rosso but also of Peruzzi.
[3] I have to thank Mr. H. E. Popham, Keeper of the Department of Prints, British Museum, for bringing this drawing to my notice. Though Ligorio signed his drawing (in which also a curtain is added as a motif to finish off the background), it does not seem to me in the least certain that his is the original composition. As for the signature, Ligorio was a notorious falsifier; moreover, the Chatsworth drawing is much more schematic than

tastic mannerist architects and painters, an archaeologist himself,[1] who lived in Rome at the same time as Girolamo. This is significant, for it shows that the late Girolamo and Ligorio, whose ancestry likewise included Peruzzi[2] and Giulio Romano, were stylistically near to each other.[3] Ligorio himself painted, probably just when Girolamo was living in Rome, a very 'archaeological' fresco, the *Dance of Salome*, in S. Giovanni Decollato.[4] This oratorium, a haunt of Florentine mannerism, with its many frescoes and altarpieces by Salviati, Vasari, Conte and Franco (we recollect Girolamo's link with one of the latter's works here) may well have been a pleasing milieu to Girolamo. Furthermore, after Girolamo left Rome Ligorio became Cardinal Ippolito's expert for the excavations in the Villa Adriana and, as we have mentioned, built the Villa d'Este for him;[5] later still he went for good to Ferrara, where he once more became Girolamo's successor, this time as the Duke's architect. Thus Girolamo—in spite of his nearness to classicism—

[1] Peruzzi had already prepared a book on the antiquities of Rome. Ligorio actually wrote one (which has its merits, despite falsifications). With him, the mannerist painter-architect of the type previously mentioned has developed into a more or less genuine scholar.

[2] Drawings by Peruzzi and Ligorio have even recently sometimes been confused.

[3] For instance, a drawing of *Augustus and the Sibyl*, in the British Museum, attributed to Girolamo, is, in my opinion, certainly by Ligorio, because of his picture of the same subject, previously in the Murray Collection, Florence.

[4] A. Modigliani ('Due affreschi di Pirro Ligorio', *Rivista dell'Istituto d'Archeologia e Storia dell'Arte*, III, 1931, pp. 184 ff.) also attributes another, equally 'archaeological', fresco in S. Giovanni Decollato, *The Beheading of the Baptist*, to Ligorio. I cannot attempt here to prove, in detail, what I believe to be certain, namely, that the composition, particularly of the figures, is by Salviati. But in the weak execution, which is obviously not by him, Ligorio may have had a hand.

[5] It was now Ligorio's turn to make drawings for Cardinal Ippolito for a series of tapestries representing the *Life of Hippolytos*, the son of Theseus, just as Pordenone was to make the cartoons for the *Labours of Hercules* for the Duke.

Girolamo's and the types have been smoothed down and are not his usual rather brutal ones. The reminiscences of Correggio and Parmigianino also point to Girolamo as the original author. Thus it is quite possible that it was Ligorio who copied Girolamo's composition. The weaknesses in both Girolamo and Ligorio (the left arms of the Christ Child and of the Madonna) make it to at least some slight extent possible that both drawings had a common model, perhaps some lost composition by Peruzzi.

appears all the more clearly as a link between Peruzzi and Ligorio in this chain of erudite, archaeologizing mannerism.

The Roman episode and Girolamo's personal reactions well illustrate his mentality and the particular stage he had reached in his development. This was a very different kind of sojourn from that in Bologna, where the artistic impressions he received and the somewhat abrupt way in which he reacted to them can easily be traced in his pictures. Apparently he painted very little in Rome. When he told Vasari how much he regretted not having studied there in his youth, he must have felt himself that his stay was belated and that his introduction to the treasures of modern art in Rome, however great his interest in them, had come too late to effect a radical change in his evolution as a painter. He lived in Rome not so much as a painter but as a sensitive, cultured intellectual, as an architect and archaeologist. His mental horizon widened, although the results of his impressions were apparently confined mostly to drawings. He greatly enlarged the intellectual and statuesque world he had created for himself in Ferrara, by absorbing the world of antiquity in Rome for which he was adapted by his original links with Ferrarese humanism. It was upon this that he mostly concentrated. His stylistic position between classicism and mannerism remained, after Rome, approximately the same as before, since he acquired from the antique suggestions in both directions and accentuated both of them (classicist construction and mannerist draperies of *Chance*).[1] But his knowledge expanded, his taste grew subtler, and his art, in his last years, became more rarefied, more sensitive than ever.

We must now return to our starting point, the Hampton Court portrait, and try to find its chronological place within Girolamo's *œuvre*. Although our knowledge of his development, particularly of its later phases, is at the moment rather scanty, one can, I believe, ascribe this likeness with great probability to his Bolognese period. We have already noticed how near are its type and expression—serious, sad, melancholy, somewhat *bel-esprit*—to those of two holy figures in his Bolognese altarpieces. In fact, the Madonna

---

[1]However, these draperies reveal not only the study of antique drapery such as that in the archaistic sculpture of the Grace which he copied, but also a knowledge of the late gothic draperies in Dürer's *Fortune*. Such are some of the sources of Girolamo's mannerist features.

of the S. Martino *Adoration* of 1531 seems an idealized version of this portrait; it shows the influence of this so much more than of Parmigianino's *Madonna della Rosa*, that one is tempted to believe them more or less contemporary. Very near also is the similarity with the rather Parmigianinesque *Portrait of the Archbishop*, which was certainly painted in Bologna in 1532 (the year of the Archbishop's visit to that town). Close, however, as these two likenesses are to each other, that of the ecclesiastical dignitary is more timid, more restrained, while the Hampton Court portrait is infused with individual expression. Colouristically, too, the latter seems more refined. The more unusual mulberry-red background takes the place of the simple, lead-grey one, just as the dress is infinitely more intricate with its glittering interplay of green, white, black, grey, gold, and blue, compared with the simple masses of violet and white of the ecclesiastical garb. The possibility that the Hampton Court portrait may have been done in Bologna is supported by Vasari's explicit statement that Girolamo was well received by the nobility of the town and painted some portraits there, very true to life. In fact, for a few years, Girolamo was apparently the fashionable portraitist in Bologna and might easily have had sitters from among the great local families (his altarpiece in S. Martino was ordered by the Boncompagni) as well as from among the many high-ranking visitors.[1] The portraits he executed in Ferrara, when he was primarily a court artist and was ordered by the Duke to paint copies of older pictures (no longer extant) of sixteen previous rulers of the Este family, are on the whole more frigid, drier, and more matter-of-fact than those made in Bologna. Social rank plays a larger part. Yet they are quite individualized and are somewhat more human than the contemporary portraits executed at the rival court of Florence. Girolamo's *Portrait of the Duchess* in Frankfurt wears the cold expression one might expect in the ruling lady of the country. In some ways, it recalls an antique marble bust with sharply cut eye-sockets. The *Man in a Beret* in Modena, traditionally called a member of the Este family, in his splendid and morose isolation, almost reminds one of a Corneille de Lyon portrait. The elegant

---

[1] Among visitors to Bologna, he painted not only the Archbishop of Pisa but, as Voss has proved, Cardinal Ippolito de' Medici (London, National Gallery, no. 20, attributed to Piombo).

young gentleman, previously in the Beit Collection, apparently a courtier, has a rather conventional smile. The Hampton Court portrait (as the costume, too, suggests) is very probably earlier. It seems to reflect something of the spirit of those intense years, the twenties and thirties,[1] that followed the easy harmony and poise of the short mundane period of the High Renaissance, and that, in its turn, was soon to give way, in large parts of Italy, to the courtly and ecclesiastical reaction with its ensuing restraint and conventionalism in portraiture.[2]

This portrait, the culmination of Girolamo's ability to individualize, can teach us as much about his art as, in their way do the anatomical drawings and the studies after the antique. It contains much of the equilibrium of classicism, a style from which Girolamo is never very far. Yet, the pure classicist artists of the early sixteenth century, such as Raphael or Sarto, who always idealize to some extent (as do, for that matter, the Giorgionesque painters in northern Italy) would never have produced such a consistently realistic and direct portrait. Strange as it may seem, only an artist who was also near to mannerism could be, if he chose, so thoroughly lifelike. One is reminded, in a negative way, of one of the *novelle* of Girolamo's friend, Giovanni Battista Giraldi,[3] which apparently records a true event and was certainly known to Girolamo. A Ferrarese nobleman brought ridicule upon himself by expressing discontent with a medal by the famous goldsmith, Giovanni Bernardi il Castelbolognese, which rendered faithfully his thin, sickly countenance, and by demanding something more healthy-looking. In the Hampton Court likeness Girolamo was certainly not unduly concerned with beautifying. It reveals, more than most mannerist portraits, what may lie behind the apparently unapproachable façade of this courtly and aristocratic art. I may be forgiven for saying that if the Hampton Court portrait did not exist, one

[1] See Appendix.

[2] Among Moroni's portraits, which are, within Northern Italy, perhaps the nearest historical parallel to those of Girolamo, only the well-known *Portrait of a Tailor* (London, National Gallery) with its very personal, melancholy expression, that can be compared with the Hampton Court portrait. It is characteristic that, already in the years just following Girolamo's activity, only persons of low social rank were allowed to show something of their feelings.

[3] *Ecatommiti* (begun in 1528, published in 1568), VII, I.

would have had to invent it. All good mannerist or near-mannerist artists conceal under the more or less abstract appearance of their compositions an extensive knowledge of inner and outward naturalism; but it is only rarely, especially in portraits, that they are prepared to make a consistent declaration of their knowledge. The Hampton Court portrait, much more than Girolamo's other likenesses, reveals a familiarity not only with the structure and surface of the face but also with its particular characteristic expression. It is this that lends it such interest within Girolamo's *œuvre*.[1]

Girolamo da Carpi was perhaps the last Ferrarese court painter of importance.[2] But to complete the picture of him, I should like to review briefly the ways in which he differed stylistically from one of his fellow countrymen, eleven years his junior, Niccolo dell'Abbate (1512-71), in whom some features of Ferrarese court art survive and who has sometimes been confused with our artist: Girolamo's portrait of Renée of France has, for instance, been attributed to him.[3]

Abbate began his career in Modena, the second largest town of the Duchy of Ferrara, which had, apparently, a somewhat freer and less rarefied atmosphere than the capital.[4] He worked for the town itself, and particularly for castles and villas belonging to the nobility in the countryside (even in

[1] Longhi's (*Ampliamenti*) encomium of the portrait of Giulia Muzzarelli, as donor in a large altarpiece, characterizing it as one of the most amazing likenesses of the sixteenth century and attributing to it a directness already foreshadowing Velázquez, could be even more aptly applied to the Hampton Court portrait. It is more natural for a praying figure, within a votive picture, to express sentiment than for an independent portrait to do so.
[2] The reservation is necessary for we know very little at the moment about the interesting painter, Bastianino (1532-1602), who worked for the last Duke, Tasso's patron, Alfonso II (1559-97).
[3] After it had first been assigned to Parmigianino, the portrait was attributed by A. Venturi to Niccolo dell'Abbate (*Storia dell'arte italiana*, IX, 6, Milan, 1939, p. 604); previously it had been ascribed by Burckhardt (tentatively) to Dosso Dossi, by Bode to Scorel, by Berenson to Sodoma.
[4] Protestantism was more widespread in Modena, both in the town itself and at the university, than in Ferrara. It was in order to meet this danger that Morone, one of the great figures of the 'early' Counter-Reformation, was appointed Bishop of Modena. Later, however, when the real Counter-Reformation was victorious, Morone was imprisoned by Paul IV and only just escaped death for his too great indulgence towards the Protestants of Modena.

Ariosto's former villa). Some of these frescoes may be described as a manner-ist counterpart of those of the international courtly gothic of the early fifteenth century. Abbate's principal work of this period, still close, if one may say so, to the spirit of the *Orlando Furioso*, are frescoes illustrating the *Aeneid* and the pastimes of the aristocracy, painted for one of the most powerful and celebrated families of the country, the Counts of Scandiano, descendants of Ariosto's predecessor in chivalresque poetry, Boiardo. In these frescoes (now in the Gallery of Modena) Abbate keeps much closer than Girolamo to Dosso's flamboyant, fantastic style, however cooled down, however elegant and mannerist it may have become.[1] In Ferrara itself, not only is much of Dosso's spirit dead by now, but even Ariosto's name no longer seems to shine with its previous lustre, at least among the erudite.[2] It is significant that when Abbate left Modena in 1548 he did not go to Ferrara

---

[1] For the similar pattern, compare, for instance, Abbate's *Descent of Aeneas and the Cumaean Sibyl into Avernus* with Dosso's *Departure of the Argonauts*, in the National Gallery in Washington. On the style of the Scandiano frescoes, see, 'A Drawing by Niccolo dell'Abbate in Windsor', *Burlington Magazine*, LXXXI, 1942, pp. 226 ff.

[2] We know that Ercole II esteemed Ariosto and that the *Orlando Furioso* remained very popular with the Ferrarese aristocracy. But, on the whole, intellectuals and lovers of antiquity now fundamentally believed that the *Orlando Furioso*, just as all chivalresque poetry, was suitable only for the ignorant and vulgar. If Giovanni Battista Giraldi tried, in his theoretical writings, to save something of chivalresque poetry at its 'best' (that is, when it was dignified and subjected to rules), it only shows what he was up against among Italian scholars.

As for Giraldi's friend, Girolamo, it is scarcely thinkable that he or anyone of his generation and mentality was any longer a consistent Ariosto illustrator. Though it is not impossible to suppose, in theory, that he occasionally painted an episode of the *Orlando Furioso*, it certainly could not have been the very fantastic picture of *Roger and Angelica* in the Kress Collection, New York, which Longhi (*op. cit.*), with a query, ascribes to him. Although, as we have amply seen, Girolamo was Ferrarese enough to reveal, occasionally, a somewhat bizarre streak, it would never have been strong enough to produce the Kress picture. (For the characteristics and limitations of the fanciful, within Ferrarese art, see the extremely different 'Flemish' landscape in the Borghese Gallery, mentioned in note 2, p. 121.) Contrary to Longhi, who emphatically asserts that the painting is Ferrarese, I believe that *this* particular variety of the fantastic, occurring about and after the middle of the century, was not possible in Ferrara but only within Florentine manner-ism. In so far as one can judge a picture the original of which one has not seen and the texture of which, therefore, one does not know, I think it is by Maso da San Friano.

but quitted the Duchy altogether for nearby Bologna, the second most important town of the Papal State.[1] Perhaps he felt that the bubbling lightness of his style was not quite suited to the intellectual Ferrarese court where Girolamo was the well-established representative, and which may have appeared to offer slender chances for secular art. At any rate, there could not have been much of a future for him in the capital, which was slowly becoming impoverished. The increasingly disproportionate luxury of the court was sucking up the wealth of the whole country[2] to such an extent that few other individual artistic commissions, even from the great aristocratic families, were any longer possible.[3] Bologna, on the other hand (though its political independence had been lost for ever), was a kind of aristocratic republic, ruled by the local families under the Papal Legate, and was on the way to becoming one of the great Italian art centres. Abbate received numerous large orders to decorate the palaces of cardinals of some literary taste, and of the aristocracy, who preferred pleasant subjects to ponderous erudition. It is in these Bolognese frescoes, among the themes of which the *Aeneid* and the *Orlando Furioso* again take prominence, that the spirit of Ariosto and Dosso still lives on, even if once again somewhat chilled. Here one is conscious not only of the pleasure taken in the fanciful, 'romantic' episodes of Virgil and Ariosto but one feels again something of the free, magic touch of Dosso, which is particularly noticeable in Abbate's charming renderings in the Palazzo Poggi of fashionable society amusements. The

[1] For a long time it had been the natural trend of the Ferrarese artists to go to the more prosperous Bologna. When Ercole II built a citadel in Modena and destroyed some suburbs for the purpose, the homeless inhabitants also went to Bologna.

[2] The process by which the court drained away the riches of the country in the interests of pomp and amusement was already going on in the fifteenth century. In the sixteenth century the economic decline of Italy, and in particular of the Duchy of Ferrara, was such that the ostentatious luxury of the court (the only means of self-advertisement at the disposal of an ambitious but politically increasingly unimportant ruler, like Alfonso II) stands out all the more blatantly. On the other hand, the cultural life of the court retained its intensity, in many fields, up to the time of its extinction in 1597.

[3] It seems to me—though this needs more careful examination—that beginning already in the thirties and forties, the main artistic commissions in Ferrara very largely derived from the Duke and that those for large altarpieces previously given by wealthy individuals were far less frequent.

frescoes for this palace were ordered by Cardinal Poggi, a member of the local aristocracy, the most influential man in Bologna after the Papal Legate. I reproduce here a broad, lively brush drawing by Abbate in Windsor, representing a *Concert* (Plate 44b), in which the Dosso touch is still very much in evidence and which constitutes perhaps one of Abbate's early thoughts for the two frescoes with musical entertainments.[1] In the frescoes of the Palazzo Poggi Abbate also continues Dosso's tradition of pure landscapes, and retains some of the 'Flemish' fantasies of Battista Dosso.[2] In some ways he is near, in others very far from, Girolamo; he lived in a different world, bolder, less sophisticated, more imaginative, and more facile[3]; his mannerism is nearer to baroque,[4] Girolamo's nearer to classicism. The difference is clearly apparent in Abbate's portraits, such as the strikingly fashionable *Man with a Parrot*, in Vienna, and the chic *Portrait of a Lady* which I discovered in the storeroom of the National Gallery, London (Plate 45),[5] the first probably, the second certainly, painted after he reached France. Parmigianinesque though they are (compare, for instance, the type of the London portrait with Parmigianino's *Amor* in Vienna), they are elegant in a more worldly and gay sense than the portraits of his model, and than those of Girolamo. There is nothing sad or resigned, nothing hard or severe in them,

[1] At the same time, the drawing seems to elaborate a more improvised sketch by Abbate in Modena, with theme and figures much alike.

[2] A. Venturi (*op. cit.*, p. 609) even wrongly attributed to Abbate the landscape imitating Patinir and Bosch, in the Borghese Gallery, which has been just as incorrectly ascribed to Girolamo (see note 2, p. 121).

[3] This perhaps explains why the Bolognese Primaticcio—who probably arranged the matter—chose him to go to the French court. And that is why Abbate, when in France, became, in his approximation to Primaticcio's style, even more Parmigianinesque and worldly than he was before.

[4] Mannerism, when consistently carried through, is probably too sophisticated and abstract a style for the illustration of Ariosto. His picturesque episodes lend themselves rather to expression in baroque, especially the very early sixteenth-century and very late eighteenth-century baroque (though he was taken up also in seventeenth-century painting in particular, and precisely in Bologna). It is not surprising that few baroque works anticipate so many of the neo-baroque pictorial features of nineteenth-century romantic painting, as do Fragonard's drawings for the *Orlando Furioso*.

[5] See 'A Portrait in the National Gallery Identified', *Burlington Magazine*, LIX, 1931, pp. 226 ff.

just as they retain, again unlike those of Girolamo, a free, painterly, Venetian technique. It is characteristic that Abbate's early *Portrait of a Woman* in Madrid already demonstrates these same fundamental differences, near though it is, in other respects, to Girolamo's portrait of the Duchess.

Of Girolamo's portraits, the one in Hampton Court, I believe, will henceforth serve us as the most human document of Girolamo's portrait painting, indeed, as one of the most human documents of Italian portrait painting of the sixteenth century. And perhaps the fact also emerges that Girolamo's works really are precious records which deserve to be sought out, understood, and assembled. One cannot claim that Girolamo is one of the very great, original, creative artists of his century. But his exquisite, intellectual, and refined art represents an important shade even within the vast wealth of this century, especially as it is a true reflection of one of the most cultured of Italian courts,[1] our knowledge of which, unfortunately, is greatly hampered by the fact that its treasures have been destroyed and dispersed, probably to an extent unequalled in any other artistic centre of Italy. Though many problems around Girolamo da Carpi still remain unsolved—it is not possible at present to make exhaustive comparative studies in scattered places—perhaps we can hope to have advanced, in some points, the understanding of his art and of its milieu.

## APPENDIX

### THE SOCIAL BACKGROUND OF ITALIAN MANNERISM

The Counter-Reformation was not alone responsible for mannerist art in Italy as is constantly stated in art-historical literature. It was not yet in being during the early part of the century, that is, precisely during the most intense phase of mannerism. The Counter-Reformation was obviously one

---

[1] I should like to mention an example to show how seriously artistic problems were taken at the Ferrarese court. Cardinal Salviati and the Cardinal of Ravenna, when present at the second performance of Giovanni Battista Giraldi's *Egle* (with Girolamo's stage decorations), expressed dissatisfaction with the 'Greek manner' of dispensing with separate acts and requested to see a further presentation in the 'Roman manner', with the division into acts. Giraldi reports this in a letter to the Duke.

of the principal spiritual forces in sixteenth-century Italy, but the deeper reason why, from the middle of the century, it was so increasingly successful and able to permeate every sphere (quite apart, of course, from the traditional loyalty of the average Italian to his Church), was the economic decline of Italy and, in particular, that of the previously wealthy and powerful Italian middle class. The decay of the Italian bourgeoisie, already foreshadowed in the second half of the fifteenth century was, as is well known, consequent upon the economic advance of other European countries, the changing of the sea routes and the resulting loss to Italy of her trade with the Orient. Hence, we find an increasing importance of the aristocracy, even revitalized feudalism, as the dominating tendency of Italian social life in the sixteenth century, most noticeable perhaps in Tuscany since the establishment of the Duchy. It is this retrogression of social process which is the real, wider background of the revival of mediaeval tendencies in Italian intellectual life, including the courtly and aristocratic, irrational and unrealistic mannerist art itself. (I cannot here go into the question of how, on the other hand, a rational Protestantism, followed later by a rational and realistic art, only became possible in countries where a strong middle class, particularly an upper middle class, existed whose interests, in every sphere, differed from the Roman Curia. And only to avoid misunderstanding do I mention that the economic and social retrogression in Italy by no means implies that in the intellectual life of the second half of the sixteenth century a great deal was not produced that would become of importance for the whole of Europe in the seventeenth century; still, modern tendencies were then often persecuted in Italy, and moreover, she had to share the rôle of forerunner with other countries far more than before.)

In the early sixteenth century there was as yet no Counter-Reformation in Italy, but a general, very sincere, deeply religious ferment was evident among the masses and part of the clergy, a parallel to the social and religious movement of the German Reformation (which itself had quite a strong influence throughout Italy, particularly among the intellectuals; most of the Italian reformists were really Protestants, though only unconsciously). In Italy the very men who were later to become the dominant personalities of the Counter-Reformation participated, at first, in this reform movement. It was some time (after reconciliation with the German Protestants had failed) before a definite borderline between Reformation and the re-action to it, the Counter-Reformation, could be drawn (when, indeed, the distinction became so clear-cut that those same personalities imprisoned their former, reform-minded colleagues). In fact, only toward the middle of the century, when the general religious ardour slowly vanished did the organized Counter-Reformation hold the field. The phenomenon, so often recurring in Italian history, of a religious agitation with social and political features cannot, of course, be said 'to begin' in the early sixteenth century. But here I can only refer, chronologically, in passing, to the religio-political upheaval in Florence of the twenties which consciously harked back, in its tenets, to the Savonarola movement of the nineties; in both movements, the middle and lower sections of the bourgeoisie, more zealous than the topmost stratum, had a large share, while the upper middle class was weak and divided. I give this short, primitive analysis merely to indicate that the intense mannerism of the early sixteenth century had, as its background, not the Counter-Reformation, as yet non-existent, but the general religious, partly also social and political, agitation of those years, just as the excited late Quattrocento gothic, the forerunner of mannerism, corresponded to the Savonarola-movement in Florence. (In other parts of Italy it corres-ponded to similar phenomena which, even if not so explosive, were yet expressive of the economic decline of Italy and of the lack of a genuine upper middle-class rule.) And just as both these upheavals contained strongly popular elements, so did the corresponding movements in art: early man-nerism and late Quattrocento gothic. And again, just as both movements possessed certain aristocratic features, so also did these styles: for popular and aristocratic art, both irrational, always show a certain kinship with each other, in contrast to rational upper middle-class art. (For detailed proof of this latter assertion, with respect to the fourteenth and early fifteenth centuries, see my book, *Florentine Painting and Its Social Background*, London, 1948.) After the democratic potentialities of the early sixteenth century were definitely snuffed out and the aristocratic tendencies reappeared, the popular traits, still to be found in early mannerism, largely disappeared, yielding place to the courtly and refined (also the scholarly). It is revealing that the mannerism of this sophisticated courtly art of the later part of the century is usually far more extreme than that of the contemporary, severe, puritan,

*160*

accurate, uncomplicated religious painting which keeps close to the spirit of the official Counter-Reformation, of the Council of Trent.

Apart from these general hints about Italian mannerism, I cannot here examine the particular social background of Girolamo da Carpi's style. At the present stage of research into Italian social history of the sixteenth century this would demand extensive study outside the framework and scope of this article. But for a complete understanding of Girolamo's style, one must, of course, penetrate beyond the mere outline of the cultural background which I have given in the text.

# 4

# Around Salviati

The portrait of a young man, in the City Art Museum, St. Louis, here
reproduced (Plate 48a) is, in my opinion, a work of Michele di Ridolfo
(1503-77); it is not by his great Florentine fellow-countryman, Salviati
(1510-63) to whom it was attributed a short time ago.[1] Though interesting
and obviously imitating Salviati,[2] it has little of the sharp characterization
of Salviati's types, but schematizes and slightly sentimentalizes them; its
somewhat melancholy, dreamy expression is a remnant, even if only a
belated, distant, almost frozen remnant of the rather languishing glance often
to be found in portraits by Michele's master and previous close collaborator,
Ridolfo Ghirlandaio—a glance which, as we know, seems to fuse Leonardo,
Piero di Cosimo and the early Pontormo.[3] The rigid triangularity of the
composition, an elegant, abstract formula similar to that of Michele's well-
known portrait of a huntsman in the Antinori collection, also greatly differs
from the sovereign freedom and variety prevailing in Salviati's portraits.[4]
Equally, the modelling of the face, flabbier, more superficial, and more
forced, is dissimilar from Salviati's heads, organically built-up with great
knowledge of structure. It is a face slightly on the slant, with long narrow
nostrils, which one also knows well though of course in a less individualized
form from Michele's rather empty-faced female half-length figures from

[1] By F. Zeri, 'Salviati e Jacopino del Conte' in *Proporzioni*, II, 1948, p. 180.
[2] C. Gamba, 'Ridolfo e Michele di Ridolfo del Ghirlandaio' in *Dedalo*, IX, 1928-29, pp.
463 ff. and pp. 544 ff. Gamba has put together Michele's *œuvre*, and has already noticed
that his portraits can, at first sight, be confused with Salviati's.
[3] See Gamba, *op. cit.*
[4] The schematic pattern of the St. Louis portrait is at least in principle much nearer to
some of Vasari's.

Roman history or mythology, his *Lucretias* or *Ledas*, say, in the Borghese Gallery. In order to provide a comparison with the face and in particular with the very characteristic nose I am here reproducing one of the less familiar of these half-length figures, a *Lucretia*, formerly on the Berlin art market (Plate 48b), a picture which, I imagine, no one would think of attributing to Salviati.[1]

The somewhat elaborate combination of drapery and landscape in the St. Louis portrait, the attempt to mass together everything desirable in the background of a likeness, is not compatible with Salviati's mentality. Apart from this, the drapery is too schematic and old-fashioned for him: centrally placed and canopy-like, it is still a relic of Sarto's *Holy Families*.[2] On the other hand, in details, it is exactly the kind of drapery favoured by Michele: compare it, for instance, with that under the left foot of *Night* in the Colonna Gallery. Furthermore, as far as I can recollect, Salviati never used landscape but usually some quiet background in his portraits. And again it is precisely this landscape of the St. Louis painting, somewhat crudely contrasted in lights and shades and very different from the delicate, differentiated, often light-toned ones of Salviati, which is characteristic of Michele, not only in the Antinori portrait but also, for example, in the paintings of *Venus* and of *Night* in the Colonna Gallery. Moreover, the river-god and the lion in the background of the St. Louis picture are too heavy and clumsy for Salviati, while the large flower is a motif for which Michele appears to have had a penchant.

To summarize: Michele, older than Salviati though outliving him, is an artist who, during his long career (at first he was a pupil of Credi, afterwards the not very original collaborator of Ridolfo Ghirlandaio, an imitator of the late Sarto and so on), always tried to follow the ruling fashion,[3] though from a certain distance and not on a very high plane. Apparently his highest

[1] Since Michele's vapid faces have some affinity with those of Vasari, this picture was originally ascribed to the latter, as so many of Michele's works. I attributed it to Michele some twenty years ago when the picture turned up in Berlin.

[2] The man's right arm is a copy after that of Michelangelo's *Lorenzo de' Medici* Statue.

[3] Gamba, *op. cit.*, rightly says that one must know Michele's whole œuvre in order to be able to distinguish his eclectic pictures, some of which even look Emilian, from those of other artists.

social level of employment was to have been brought in by Vasari for the decoration of the Palazzo Vecchio, though, as is obvious from the latter's condescending remark, not upon very prominent tasks. At a date not very distant Michele created the St. Louis portrait in the manner of Salviati, the fashionable model for the painters working round Vasari for the court, yet a portrait ultimately very different from him. While Salviati, a painter of great originality and intensity in his mannerist figure compositions, not only consistently displays an extensive naturalistic knowledge in his portraits,[1] but also represents his sitters in natural poses, Michele, in trying to be up-to-date, appears, in his portraits, with their slantwise construction and gliding motions (in the manner of Parmigianino's *Madonnas*) more mannerist than Salviati. However, this peculiar intensity becomes in him, a lesser artist, somewhat conventionalized, while the chief background motif, the showy drapery, is particularly out-of-date. Of course, the form that out-modedness takes in an artist, striving very hard to be the opposite, is always interesting and revealing. The unusual combination of a lingering Ridolfo Ghirlandaio-like sentiment, characteristic of an older generation and of bygone-decades, with a schematic, grand mannerist framework, lends this likeness, composite, backward and fundamentally second-rate as it is, an individual flavour (one could almost say attraction), naturally absent from other more unequivocal portraits by greater artists of the period, like Bronzino or Salviati.

Research on Salviati has not made great progress in recent years, not even the mere clarification of his *œuvre*. On the contrary. While, as we have just seen, a portrait by Michele has been attributed to him, one of his most important paintings, the *Three Fates* in the Pitti has been assigned to Jacopino del Conte.[2] Another though slightly unusual picture by him, the *Nativity*, likewise in the Pitti, has even more recently been given to Girolamo da Carpi (with a query).[3] My simple assertion that these two latter pictures are

---

[1] See, on this aspect of portraits by some mannerist artists, my 'Observations on Girolamo da Carpi', p. 107.

[2] By Zeri, *op. cit.*

[3] By R. Salvini, *Mostra di Lelio Orsi*, Reggio Emilia, 1950, p. 68. Voss's correct attribution (*Malerei der Spätrenaissance in Rom und Florenz*, Berlin, 1920, I, p. 248) to Salviati is not alluded to, even in the bibliography. Incidentally, in the above-mentioned catalogue, Salvini also retains (though with some doubt) the attribution to Orsi of the *Capture of*

## Around Salviati

by Salviati cannot, of course, in itself clear the air, since the question should be treated at length, with detailed arguments and copious illustration. However, in the present rather chaotic situation of Italian mannerist studies, even these crude statements may serve as a useful *point d'appui* for someone who attempts the overdue task of bringing together Salviati's *œuvre* and of seriously working out his stylistic development. So, too, the differentiation of Michele di Ridolfo from Salviati may prove helpful for the equally very interesting work of elucidating the taste of the wide public that favoured Michele (he had a large workshop, and even a very close imitator in Brina) and bought his oft-repeated devotional pictures and mythological female figures.

---

*Christ* in the Oratorio del Gonfalone. This fresco, however, has nothing to do with Orsi; it is influenced by Muziano, is almost certainly by Nebbia and thus probably painted some twenty years after Orsi's stay in Rome.

# 5

# Mr. Oldham and his Guests by Highmore

The picture here reproduced has recently (1948) been acquired by the Tate Gallery, where it is styled *British School, 18th Century: Conversation Piece* (Plate 46). But it is easy to go beyond this. Ample testimony exists as to the painter of the picture, the people represented in it and the occasion on which it was painted.

In his well-known *Nollekens and his Times*,[1] J. T. Smith, Keeper of the Prints and Drawings of the British Museum in the early nineteenth century, discussed the picture at length; according to him, it was wrongly attributed to Hogarth in the 1827 sale catalogue of the pictures of the Rev. Theodore Williams, where the figures were equally erroneously stated to be Hogarth, Dr. Monsey, and Old Slaughter, in whose celebrated coffee-house the conversation over a bowl of punch was said to have taken place.[2] In fact, there was every reason why Smith, who had been informed of its sudden cropping up at this sale by his friend Lewis, should have known the picture intimately since his youth. I quote Smith himself:

> To my great surprise [I] found it to be a picture that had been for the first eleven years of my life in my sleeping-room, and it gives me no small gratification to state that this picture, so roundly asserted to be from the pencil of Hogarth, was produced by Mr. Highmore. I agree with Mr. Lewis as to its being wonderfully well painted; indeed, it is equal, in my opinion, to many productions of Hogarth in the portrait way: but the picture was painted by Highmore, for Nathaniel Oldham, my father's godfather and one of the artist's patrons. . . . My father's account of this picture was that Mr. Oldham had invited three friends to dine with him at his house at Ealing; but being a

[1] London, 1829, pp. 219 ff.
[2] That the two main personages in the foreground were called Captain Coram and Bishop Hoadly only shortly before the picture's acquisition by the Tate Gallery, is also a relic of the old, persistent attribution to Hogarth.

famous and constant sportsman he did not arrive till they had dined; and then he found them so comfortably seated with their pipes over a bowl of negus, that he commissioned Highmore to paint the scene and desired that he might be introduced in it just as he then appeared. A man on the right, with a white wig and black coat, was an old schoolmaster; and one opposite to him a farmer, both of Ealing; another, in the middle, in a red cap, was the artist, Highmore; and one with his hat on, behind the farmer's chair was Nathaniel Oldham. When Mr. Oldham died, his property was sold; but this and one or two other family pictures were given to a relative, from whom father purchased it, as it contained the portrait of his godfather. It afterwards became the property of Mr. Bellamy, a linen draper, residing in Queen-street by the Mansion-house.

There can be little doubt that Smith is describing precisely the picture now in the Tate Gallery. Furthermore, added to the identity of the theme, on the basis of Highmore's *Self-Portrait* in the Gallery at Melbourne (Plate 47a), it is evident that the artist portrayed in our picture sitting between the farmer and the schoolmaster is Highmore himself (though now somewhat older). And the man in the left corner, equally without any doubt, can be identified as Mr. Oldham, a wealthy gentleman and collector of curiosities and paintings, whose appearance we know from another picture of him (somewhat more youthful) in a hunting coat and holding a gun, by his friend Highmore, after which there exists a mezzotint by J. Faber, of about 1740.[1]

And what about stylistic confirmation? It is, of course, easy to be wise after the event, that is, after having read the passage in question in Smith, but it is certain that the picture fits extremely well within Highmore's *œuvre*. The quiet, unobtrusive way of arrangement, the massive, sharply shadowed but slightly spongy faces,[2] the particular, restrained sober colours, the rather schematic, yet pronounced treatment of light and even the dashes of vermilion in the nostrils and between the fingers are characteristic of this artist. Although we are dealing here with a portrait group, we find in other of

[1] In the literature, we constantly find an indication that this mezzotint is from 1740. Yet the engraving is, in fact, undated. In the copy in the British Museum Print Room of J. C. Smith's *British Mezzotinto Portraits* (London, 1883) is inscribed a marginal note to the effect that the costumes in the engraving are of 1740.

[2] Although Hogarth is by far the more baroque artist of the two, his figures and faces are more solidly constructed than those of Highmore.

# Mr. Oldham and his Guests by Highmore

Highmore's paintings similar types of expression, notably the schoolmaster's, which is reminiscent of various figures of the *Pamela* series or of the bewigged gentleman in the right foreground of his *Green Room in Drury Lane* (Lord Glenconner's Collection).

What is more, our painting gives us much needed additional elucidation on important features of Highmore's art.[1] The picture arose from an entirely casual occasion, typical of the new, easy-going, homely atmosphere of middle-class society. The people represented, in particular the two shown most prominently, the farmer and the schoolmaster, are average personages of no rank, who behave naturally and informally. Far from feeling embarrassed as newcomers to the ranks of sitter, the farmer and the schoolmaster appear very self-possessed and all the more impressive since the picture, containing half-length figures of only four people, is relatively large ($41\frac{1}{2}$ by 51 in.). All four of them are very genuine, intentional character studies, such as we do not often meet in English art of that time, apart, of course, from Hogarth: the jovial farmer, with his sense of well-being; the morose schoolmaster, who ill conceals his liking for drink behind a pseudo-pomposity; Highmore, smiling slyly if somewhat subdued by the farmer's bulky heartiness; and Mr. Oldham himself, rather in the background as becomes the host, leaning nonchalantly on the farmer's chair, while thoroughly enjoying the whole situation. The painting has nothing of the baroque ceremonial and properties of aristocratic portrait groups. Indeed, if we take it together with some of his other works, we can detect in Highmore the germs of a realistic classicism, most unusual in the England of his time, where the baroque tradition was still overwhelmingly strong. In our Conversation Piece, from the unaffected, naturalistic attitudes there grows a massive, quiet composition, roughly in verticals and straight diagonals,

[1] On the whole, Highmore's art is so little known and appreciated that all his outstanding paintings have been, on first thoughts, attributed to others: his *Pamela* cycle, one of the most important works of the English eighteenth century, which is so distressingly split up between London, Cambridge and Melbourne, was assigned to Troost, the *Green Room* and our Conversation Piece to Hogarth. Most appreciation of Highmore's qualities is to be found in the various writings of C. H. Collins Baker, to whom is also due the attribution to our artist of the *Green Room* and of the very interesting Conversation Piece in the collection of Lord Plymouth (previously given first to Hogarth, then to Jonathan Richardson).

and to this simple but well-balanced design corresponds a relatively well-calculated harmonizing of the sparing colours. The brighter and more lively colour spots of the two middle figures (fawn of the farmer, wine-red of Highmore) and of the still-life on the table are framed, with skilful transitions and echoes (faces, wigs, shirt-cuffs, etc.) by the blackish-grey of the teacher, the more brownish-grey of Mr. Oldham (his black hat) and the dark-grey background, so that the whole constitutes (perhaps with the exception of the slightly raw colouring of the farmer's face) an economical well-graduated, though not exciting, colour pattern.

Highmore is an artist of much smaller stature than Hogarth, yet he is in himself interesting and, with his different style, equally significant of that great period of English middle-class art, the first half of the eighteenth century.[1] Hence a comparison with Hogarth greatly helps to put into relief, even if only in a negative sense, the characteristics of our picture or, for that matter, of much of Highmore's *œuvre*. Beside Hogarth's most varied, flickering baroque-rococo compositions, the originality and exuberance of his colours, Highmore's pictures look rather schematic, almost drab. Compare, for instance, his paintings of the *Pamela* series with the contemporary *Marriage à la Mode*. While Hogarth's pictures seem to be iridescent in their innumerable details and are held together, as it were, only in the final resort,[2] Highmore's compositions are unified not only through their scanty colours but most obviously also by simple, summarizing lines. Or, once again, to put it more positively: there is a tendency in Highmore—and some of his *Pamela* pictures, with their full-length figures shown in action, are an even better example than our Conversation Piece—towards a realistic classicism. This, as we are well aware, is the progressive art of a developed middle class on the Continent, an art precariously balanced between an overflowing

---

[1] Highmore is characterized by his contemporaries as a steadfast, sober man, much interested in the scientific side of his art, particularly in anatomy and perspective. From every angle we gain the impression of a solid member of the middle class, which indeed he was, not only in his mentality but even through his birth and upbringing: he was the son of a coal merchant, was educated at the Merchant Taylor's School, at first studied law and became an articled clerk.

[2] See, on Hogarth's way of composition and its origins, 'Hogarth and his Borrowings', *Art Bulletin*, XXIX, 1947.

realism, on the one hand, an empty classicism on the other—an art, therefore, which needs a long tradition to crystallize. Even if a bourgeois artistic
tradition is missing in England (on account of the previous hostility of
Puritanism to the visual arts), it is very important that, in the heyday of
English middle-class culture, at the time of Richardson and Fielding, there
should have existed in art such an advanced tendency, or, at least, clear
suggestions of such a tendency. Though Highmore's artistic level, needless
to say, is much lower than that of those two great artists, in some ways, he
fits—I think, for instance, of his *Mr B finds Pamela Writing a Letter*—into a
line of stylistic evolution which leads from Vermeer to Chardin: the line
of a realistic classicism, of scenes of an intimate, domestic character, of a
simple but calculated figure pattern, of a reduced but refined colour scheme.
In fact, it is no surprise to learn that Highmore studied not only in Holland
and Flanders but probably also in Paris and it is difficult to believe that he
could have been entirely unacquainted with Chardin.[1]

No thorough chronology of Highmore's pictures has ever been attempted
and, in consequence, an exact dating of the Tate picture, within the framework of his whole development, cannot yet be undertaken. But there is
sufficient external and internal evidence to show that it is a rather late painting, as are, apparently, most of Highmore's important works. Highmore
(born in 1692) looks about fifty to sixty years old, perhaps nearer to fifty.
The costumes equally point towards the 1740s and even 1750.[2] As to Mr.
Oldham's personal dates, we are less informed; we know that he acquired
the house in Ealing, where the scene takes place, as early as 1728, but we do
not know when, having lost his fortune, he had to leave it, nor do we know
the date of his death (he allegedly died in the debtors' prison)[3]; so we have

[1] Highmore's journey to Paris is alleged to have taken place in 1734. As we know,
Hogarth, who was in Paris in 1743 when Chardin's fame was greater, much appreciated
the French artist.

[2] I am grateful for this suggestion to Sir Henry Hake and to Mr. C. K. Adams, of the
National Portrait Gallery.

[3] J. C. Smith (*op. cit.*) says that Oldham died in 1740, but I do not know whether this date
has any foundation in fact. J. Caulfield (*Portraits, Memoirs and Characters of Remarkable
Persons*, London, 1819) does not mention the date of his death nor does the article in the
*Dictionary of National Biography*. Nathaniel Smith (later assistant of Roubiliac and father
of J. T. Smith) to whom Mr. Oldham was godfather, was born in 1740 or 1741.

no dates to serve us, from this factual point of view, as *ante quem*. However, there seems to me also obvious stylistic proof for the late date. Highmore's early portraits incline towards baroque and rococo, the later ones towards a realistic classicism. From the start, he painted portraits of the aristocracy as well as of the wealthy middle class, and it can be said, by and large, that in his early phase, when the English bourgeoisie had as yet no artistic tradition, likenesses of both types of sitters follow the elegant Kneller and the French-courtly tradition, give rather generalizing facial features and ostentatious attitudes and gestures. Later, as the middle class and its new, quite unique culture steadily grew in influence, a realistic portraiture, too, slowly evolved. We now often find in Highmore an organic mixture of graciousness and simplicity. At the same time, he seems to incline to a certain dualism of style, according to the sitter, adhering, in those of the middle class, to a more true-to-nature portrayal of individualized features, a plain, quiet monumentality, combined with a deliberate attempt at clair-obscur; perhaps, in this late phase, even his portraits of the aristocracy tend to show some influence of the other type.[1] Indeed, *Mr Oldham and his Guests* seems to be the climax of Highmore's evolved, late realism, and it is interesting to compare it with the delicate realistic classicism of his portrait of 1750—monumental in spite of its small size—of his friend, the author of *Pamela* (National Portrait Gallery, Plate 47b).[2] It is as if the rather typified figures of his *Pamela* illustrations, painted about 1744, had now acquired real, pulsating life in this casual and genuine meeting of friends, where Highmore gives character studies, more sharply observed than in any other of his works.

[1] In its very general outlines, the development of Hogarth's portraiture is somewhat similar, but even this can be said only with reservations. While in Hogarth, the baroque-rococo trend is more pronounced than in Highmore, yet, according to the public for which he worked, his range of style is wider and hence there are greater extremes within his *œuvre* (this is true not only of his portraits). Among Hogarth's portraits, there appears to have been, as can perhaps be assumed from a survey of the rich material available in Sir Robert Witt's library, a slight accent upon sitters from elegant society and the aristocracy; among those of Highmore, upon rich city merchants. As is also well known, it was Hogarth who introduced into England early in his career as a painter, the small-scale portrait groups of the aristocracy, infusing them with French-rococo graciousness.
[2] With the affinities between Highmore and Richardson, on the one hand, Hogarth and Fielding on the other, I am dealing, in detail, in my forthcoming book on Hogarth and his place in European art [published 1962, Ed.].

# Mr. Oldham and his Guests by Highmore

*Mr Oldham and his Guests* is a portrait group which could not have been painted at that time in any other country, not even in the France of Chardin, which was too well educated and perhaps not intense enough. If we wish to find, in France, something similar to it or to some robust Hogarth likenesses, such as those of his mother (David Rothschild Collection) or *Mr Arnold* (Fitzwilliam Museum), it is to a later development of the French middle class, to the late eighteenth and early nineteenth century, to David and his school, that we must look, to a time when, in the stream of history, the keen, very early English middle-class culture of the first part of the eighteenth century bore its fruits on the Continent.[1] It was in Highmore's and Hogarth's England that those new renderings of bourgeois self-confidence, those very individualized expressions, those informal poses originated.

[1] The solid, direct portrait of *Mr Arnold*, probably painted shortly before Highmore's portrait group, closely resembles the portly farmer of this picture, while it even foreshadows Ingres (David's pupil), his portrait of *M. Bertin* (1832, Louvre), the rich newspaper proprietor, which is known—to quote a phrase current even in handbooks—as the most typical portrait of a self-confident member of the middle class.

# 6

# *Remarks on the method of art-history*

The following are a few casual thoughts, in no sense systematic, on the method of art-history, which have occurred to me while looking through some art-historical literature of the past years.

It is, of course, platitudinous to say that art-history deals with the history of art, that it combines and connects art and history. It is equally obvious that the method used in art-history, as in other disciplines, undergoes certain changes from generation to generation. That of each generation depends on how it views art and how it views history and on the differing combination and proportion of the two components which, as a result, arise afresh in every generation.[1] So the method of art-history naturally constitutes a part of the prevailing intellectual outlook, the problems and interests, of successive periods. Alterations in art-historical methods do not in the least cancel out achievements of previous generations, but only effect a shift of accent which brings into relief ideas in art, as in history, which the particular generation considers most important. For not only do the various methods differ in the importance they accord to history, but they are also largely determined by the preoccupations of historical research itself in the period in question.

Compared with earlier methods (say, with Karl Justi, who described the cultural background and the personal character of great artists), Wölfflin's formalistic method conceded, relatively, the smallest place to history. His approach, to a greater degree than that of his predecessors, tended ultimately

[1] This, of course, is over-simplification. It was particularly during the heroic years around 1900, spiritually so rich and complex, that various methods of art-history, to a certain extent, over-lapped. However, seen in perspective, the main trend of development is clearly discernible.

to reflect the then prevailing doctrine of art for art's sake. This thesis, as is well known, had been conceived by a group of French Romantics and propagated mainly in the fifties and sixties of the nineteenth-century by writers and poets, who believed in erecting an ivory tower for themselves, who considered art to be detached from the ideas of their time, and who stressed in art the 'eternal', the 'absolute', that is, the purely formal values.[1] Wölfflin's very lucid, formal analyses, behind which is an undisguised bias in favour of the classicist Cinquecento composition, reduced the wealth of historical evolution to a few fundamental categories, a few typified schemes. The Viennese school of art-history, to which Riegl, Wickhoff, and Dvořák belonged, gave a far more prominent place than did Wölfflin to history and the historical development of style.[2] Here, works of art were treated as threads in the stylistic development, and so great was the value placed upon the continuity of this evolution that so-called 'dark periods', 'periods of decay', like those of the late antique, of mannerism, of baroque, that is, periods of which previous art-historians had disapproved, were no longer recognized as such, but were studied constructively and with particular thoroughness. Although scholars of the Viennese school made most exact, formal analyses, even as inexorably logical as Riegl's, they—the late Riegl himself and particularly Dvořák—combined them with analyses of themes and of thematic features. Continuing Riegl, who, in his late phase, regarded his notion of the 'art-will' as dependent on the outlook of the period in question, Dvořák, in his later years, dealt with art-history as part of the history of ideas, of the development of the human spirit. As I was myself a pupil first of Wölfflin and then of Dvořák, I can still feel the great difference in the spiritual atmosphere of these two scholars. I should like to mention a

---

[1] In A. Cassagne's well-documented book, *La Théorie de l'Art pour l'Art en France*, Paris, 1906, we read how this theory, originated by Théophile Gautier, developed and under what social and historical circumstances it finally got the upper hand, in spite of early resistance from Victor Hugo and George Sand.

[2] It is no mere chance that, at the University of Vienna, the Art-Historical Institute took its place within the framework of the Austrian Institute of Historical Research. In his articles on Riegl and Wickhoff, Dvořák describes the struggle of both these scholars against aesthetic dogmatism, and characterizes Riegl's method as the victory of the psychological and historical conception of art-history over an absolute aesthetics.

characteristic example of the wide-embracing scope of Dvořák's approach. When writing of the art of the Van Eycks as early as 1904 he remarked that art-history had so far offered no explanation of its sudden emergence, but that the exploration of the sources of the new bourgeois culture in Flanders, of which this art was a product, could only be found in books of economic history. Wölfflin would never have said anything approaching this.

However, it has been chiefly in recent decades since Dvořák's death, as a more general interest has been taken in economic and social questions, that economic and social history within history has made such rapid strides— parallel to the sudden rise of sociology and the social sciences. It was almost twenty years after writing his *History of England* that G. M. Trevelyan gave us, in 1942—a sign of the new trend—his *English Social History*. How the history of ideas, which previously led a comparatively isolated existence, has come to be closely connected with social history, so that certain types of outlook, in a given period, take on a clear outline, is well seen in R. H. Tawney's *Religion and the Rise of Capitalism* (London, 1926), a justly re- nowned example of this new tendency. The importance of social and religious history for an understanding of the history of literature and the entirely new interpretations resulting from it are shown—to name one book among many—by G. Thomson's *Æschylus and Athens* (London, 1941).[1] The favourite field of art-history, the Italian Renaissance, has lately been worked through, from the new angle, in A. v. Martin's *Sociology of the Italian Re- naissance* (English translation, London, 1944). Like the other historical sciences, the history of religion or the history of literature, art-history too is now taking notice of, and using for its own purposes, the ever closer co-operation between the various historical disciplines and the broadening that has taken place in historical research through a mounting interest in social history. All the more, since our views not only on history but also on art have been modified. We have come to look at art, just as history, in a less esoteric

---

[1] Some forgotten but valuable books have now become topical for the same reason. In consequence of the interest recently taken in social analyses of the literary public, A. Beljame's book of 1881, *Le Public et les Hommes de Lettres en Angleterre au 18e siècle*, has just been translated and published in English. (Editor's note: London, 1948.)

light, associated more closely, in devious ways, with problems of real, every-day life; hence, for instance, the increasing attention given to the subject-matter of works of art—a clear indication that the art for art's sake point of view has much weakened.[1] It is this new combination of the two components which characterizes the method of art-history in our generation.

Here it was Warburg, with his wide range of interest in many cultural and historical disciplines, who did most of the pioneering work and whose life-long activity so clearly contained the germs of a new method of art-history. I will confine myself here to recalling his numerous well-known essays, between 1902 and 1907, devoted to the examination of the mentality and artistic taste of the Florentine middle-class patrons at the time of Lorenzo de' Medici. Since his death his research work has been continued in the same spirit by the Institute which bears his name and which is now incorporated into London University. Warburg's point of view is best summarized in the words of his own disciples. In her introduction to War-burg's writings, Dr. Gertrude Bing describes how, aided by material in the Florentine archives, he succeeded in rescuing the work of art from the isolation with which it was threatened by a purely aesthetic and formal approach. In examining in each case the inter-dependence between the pictorial and literary evidence, the relation of the artist to the patron, the close connection between the work of art, its social milieu and its practical purpose, Warburg took into consideration not only the products of great art but also minor and aesthetically insignificant works of pictorial art. Or, to use the terminology of another scholar associated with the school of Warburg, E. Wind[2]: Warburg was just as averse to the autonomy of a Wölfflinian, isolationist art-history as to the artificial boundaries between the 'purely artistic' and the 'non-artistic' factors, constructed by art-historians. In fact, works of popular and half-popular art were, and are, constantly adduced by Warburg himself and by scholars of the Warburg Institute, in

[1] That is why—to take an outstanding example—such a widespread interest is now shown among the public in Hogarth, who, not many years ago, was still looked down upon in art-history and dubbed a 'literary' artist.

[2] See his introduction to the *Bibliography on the Survival of the Classics*, edited by the Warburg Institute, London, 1934.

particular by the late F. Saxl, for an understanding of the whole art and the whole world of thought of a period.[1]

The severely historical spirit of the school of Vienna and the resolutely anti-art-for-art's-sake attitude of Warburg together paved the way for a deeper, richer and less nebulous study of art-history, which can draw upon the very tangible results of the historical disciplines, in particular of social and economic, of political and religious history (not exclusively of the history of literature and philosophy) as well as of an historically-intentioned social psychology. Art-historians are now in a position to take seriously into account the many-sidedness of any one period, the complexity of types of outlook, and the mode of thought among various sections of the public,[2] in order to discover which style belongs to which outlook on life—the notion of style, of course, not being restricted to formal features, but including subject-matter. If we look at the whole of society, not only its topmost layers, we come to understand the *raison d'être* of all pictures, not

[1] How little Saxl cared for the 'boundaries' of art-history, is shown, to take one instance, in his article 'The Classical Inscription in Renaissance Art and Politics' (*Journal of the Warburg and Courtauld Institutes*, IV, 1940-41), where he has treated together copies made by humanists of ancient inscriptions and of ancient monuments, stressing the political implications of the former for the men of the Renaissance.

As is well known, scholars of the Warburg Institute have often been able, by means of an historical approach, to explain the subject-matter and to re-create the real meaning and spirit of works of art which previously had been entirely misinterpreted by generations of writers. In the case of Botticelli's mythological pictures, this has just been rectified by E. Gombrich ('Botticelli's Mythologies: A Study in the Neoplatonic Symbolism of his Circle', *Journal of the Warburg and Courtauld Institutes*, VIII, 1945), in that of Mantegna's by Wind (*Bellini's Feast of the Gods*, Cambridge, Mass., 1948). On this occasion, Gombrich writes: 'The beautiful pages which have been written by masters of prose on the emotional import of Botticelli's figures remain purely subjective unless the context in which these figures stand can be established by outside means', and Wind: 'Mantegna's *Parnassus* has had the singular misfortune of being praised for the very qualities which it attempts to mock'.

[2] What G. M. Trevelyan writes of England is true of all countries: 'In everything the old overlaps the new—in religion, in thought, in family custom. There is never any clear cut; there is no single moment when all Englishmen adopt new ways of life and thought. . . . To obtain a true picture of any period, both the old and the new elements must be borne in mind' (*English Social History*).

only the best, the most famous, the full meaning of which cannot, indeed, be really grasped in isolation. The more carefully it is scrutinized, the more easily and naturally does the social, intellectual, and artistic picture throughout a period slowly unfold itself and the way in which its parts are connected become increasingly clarified. This, then, is the kind of process now taking place in art-historical literature, particularly, but by no means exclusively, in America.

The various authors represent very different individual shades and manners of approach, yet, historically speaking, they all form part of one trend. I cannot, of course, list the multiplicity of themes which have been examined of recent years in closest connection with the actual life and thought of different periods. But readers would, I think, like to cast a rapid glance at a few suggestive examples.

Herbert Read, treating the function of art in society, has explored the general nature of the links between the forms of society at any given period and the forms of contemporary art.[1] R. Krautheimer has shown that the purely formal approach of recent times to mediaeval architecture had entirely obscured the elements which, in the view of mediaeval men, were outstanding in an edifice: namely, its religious implications, that is, its 'content.'[2] M. Schapiro's numerous writings have also thrown completely new light on certain aspects of the art of the Middle Ages: he has associated, for instance, the style of the Ruthwell Cross of seventh-century Northumbria, or the differences between the Mozarabic and the Romanesque styles practised concurrently at the end of the eleventh century in the monastery of Silos in Castille, with the religious struggles and the social and political transformations of those times.[3] For the past decade or so, an ever-increasing

---

[1] *Art and Society*, London, 1947.

[2] 'Introduction to an "Iconography of Medieval Architecture",' *Journal of the Warburg and Courtauld Institutes*, IV, 1940-41.

[3] In Northumbria, the struggle took place between the Celtic, particularist, monastic Church, shaped by the conditions of tribal society and the Roman Church, which was aiming at the integration of local peoples into the larger ambient of European and Mediterranean life ('The Religious Meaning of the Ruthwell Cross', *Art Bulletin*, XXVI, 1944); in his other article ('From Mozarabic to Romanesque in Silos', *Art Bulletin*, XXI, 1939) Schapiro has explained the coexistence of the fantastic conservative, with the more

literature has been appearing on the working conditions of artists of the Italian Renaissance, particularly in Florence, on their position within the guilds, on the various kinds of commissions, on patronage, on the prices received, etc.[1] Above all, we begin to see more clearly than before how the various styles within Italian art of this period were deeply rooted in the types of outlook and in the social and political conditions of the period.[2] M. Meiss, for instance, when enumerating the characteristics of Tuscan painting in the second half of the Trecento—abandonment of three-dimensionality and of perspective, limitation of the movements of figures, contrasting colours, ascetic or emotional expressions—has defined them as expressing a state of mind influenced by the economic crisis beginning in the forties and by the shift of power from the merchants and bankers to the lesser guilds and the lower middle class, bearers of a more conservative culture.[3] It is worth mentioning that, working independently through the same historical sources and the same literature of social history, I came to identical results myself, contrasting the Florentine painting of this period with the realist classicism of the early fourteenth (Giotto) and early fifteenth centuries (Masaccio) when the more rationalist upper middle class was in power.[4]

---

[1] A. Blunt has shown that the struggle of the artists to better their social position decisively influenced their art theories (*Artistic Theory in Italy* 1450-1600, Oxford, 1940).

[2] Warburg's friend, the economic historian, Doren, already saw, half a century ago, the emptiness of un-historical discussions on Florentine art. In his book of 1901 on the Florentine woollen industry of the fourteenth and fifteenth centuries, a standard work on this period, he affirms that the knowledge of social and economic history of that time would dispel for ever the conception of Florence as a community living in conditions of carefree prosperity, general harmony, and timeless beauty.

[3] 'Italian Primitives in Konopiste', *Art Bulletin*, xxviii, 1946.

[4] *Florentine Painting and its Social Background*, London, 1948.

---

naturalistic modern, style as due to the steady change then occurring in the outlook of the increasingly centralized Spanish Church and ultimately to the transition of Christian Spain from scattered agricultural communities to powerful centralized states with urban secular middle classes. Dr. Joan Evans' *Art in Medieval France: A Study in Patronage* has just been published by the Oxford University Press (Editor's note: London, 1948); she shows French mediaeval art as the mirror of society for which it was produced, by explaining that this art took the forms it did because of the needs of the different sections of society who commissioned it.

# Remarks on the method of art-history

E. Gombrich, having demonstrated how Botticelli's mythological pictures are firmly rooted in the literary and philosophical outlook of Lorenzo di Pierfrancesco de' Medici's circle, suggests an important parallel between the different political views of Lorenzo il Magnifico and Lorenzo di Pierfrancesco and their differing artistic tastes: Ghirlandaio and Bertoldo in contrast to Botticelli.[1] A. Blunt has sketched the connection between the social and political events, the mode of thought and the artistic theories in fifteenth- and sixteenth-century Italy.[2] In an article on Greco's so-called *Dream of Philip II*, the same author derives the formal features from the complex thematic elements, theological (Adoration of the Holy Name of Jesus) as well as political (Holy League of the Papacy, Spain, and Venice).[3] Again, in his book on Mansart, Blunt points out how the somewhat romantic classicism of this architect was suited to the court aristocracy and the rich financiers imitating them, for whom he worked[4]; and further, how the style of Mansart's churches differs according to the particular type of religious belief of the Order in question.[5] Saxl has equally sought the explanation of Aniello Falcone's realistic battle pictures which contain no specific hero, in the social type and taste of the particular Neapolitan patrons of this artist in

[1] *Op. cit.*

[2] Alberti's rational art theory, Blunt finds (*op. cit.*), was dependent upon his political outlook, that of the pre-Medici Florentine city-state, while the mystical Neoplatonic art theory was suited to the state of mind prevailing during the Medici autocracy. He asserts that the irrational, neo-mediaeval tendencies of mannerism and mannerist art theory, are only comprehensible against the background of political and religious reaction caused by the destruction of the great merchant republics with which the Papacy had been allied and by the Papacy's move from a leading place among the progressive states of Italy to one of reaction, subsequently allied with an almost feudal Spain.

[3] 'El Greco's "Dream of Philip II": An Allegory of the Holy League', *Journal of the Warburg and Courtauld Institutes*, III, 1939-40.

[4] He explains its difference from the severe classicism of Poussin and Corneille, who express the progressive and earnest ideals of civil servants and of the merchants of Paris and Lyons (*Mansart*, London, The Warburg Institute, 1941).

[5] In his book on the artistic theories in Italy, Blunt has shown that the worldly, emotional religion of the Jesuits preferred the emotional, pre-baroque tendencies in mannerist painting (Barocci).

the second-third of the seventeenth century.[1] Wind has demonstrated that Reynolds' grand solemn style and Gainsborough's simple, natural style corresponded to the two types of outlook then prevailing: the first to the heroic nature of Dr. Johnson's and Beattie's attitude, the second to the human and sceptical conception of Hume.[2] In another of his writings the same author has shown how a new trend in history painting, based on an accurate rendering of contemporary events, drew its impulse from the democratic ideas proclaimed by the American artists, West and Copley, at the time of the War of Independence; further, he makes revealing comparisons between the styles of history painting as they arose from the American and French Revolutions.[3] Schapiro has indicated how the discovery and appreciation of the folk art of the lower classes took place in a circle of radical artists and writers, among them Courbet, who sympathized with the Revolution of 1848, and how a knowledge of this art had a definite bearing upon Courbet's realism.[4] 'Backward' pictures, even of recent epochs, are now considered to be interesting and worth explaining on account of the

[1] Wealthy gentry and cool-headed businessmen, not warrior types nor politicians but closely associated with and affected by warfare and civil strife (Masaniello) and having a preference for violent and descriptive realistic art, such as was produced in various parts of Europe (The Battle Scene without a Hero: Aniello Falcone and his Patrons', *Journal of the Warburg and Courtauld Institutes*, III, 1939-40).

[2] 'Humanitätsidee und heroisiertes Porträt in der englischen Kultur des 18. Jahrhunderts', *Vorträge der Bibliothek Warburg*, IX, 1930-31.

[3] 'The Revolution of History Painting', *Journal of the Warburg and Courtauld Institutes*, II, 1938-39. In his *English Expressionist Artists in the 19th Century* (Thesis at the Courtauld Institute, 1938), E. M. Zwanenberg-Phillips bases his explanation of Blake and his followers upon an analysis of the social background.

[4] Schapiro has further noted how the difference in the social and political constellations existing before and under Napoleon III caused Courbet's friend, Champfleury, who had also belonged to this circle, to give different interpretations of popular art during the two periods ('Courbet and Popular Imagery', *Journal of the Warburg and Courtauld Institutes*, IV, 1940). The re-discovery of the le Nains by Champfleury ('The Revival of the le Nains', *Art Bulletin*, XXIV, 1942) and that of Vermeer by Bürger-Thoré, when a political exile under Napoleon III, have equally been shown by S. Meltzoff to be a result of the predilection for realism of the same circle, whose aesthetics were influenced by their democratic ideas ('The Rediscovery of Vermeer', *Marsyas*, II, 1942, New York University).

particular outlook they represent. For instance, in 1938, two exhibitions were organized at the Baltimore Museum of Art and the Walters Art Gallery: one centring round Courbet, the other displaying his contemporary adversaries, the academic counter movement; the explanatory lectures by members of different faculties of Baltimore University, later published, discussed at length the point of view not only of the naturalists, but also of the conservative official artists of the Second Empire.[1] And finally, to include a work which deals with modern art, S. Giedion has examined the relation between architecture and social development in Europe, particularly in London, and in America in the nineteenth and twentieth centuries.[2]

To acknowledge the significance of social development and of different types of outlook for understanding the diversity of styles and stylistic evolution does not, of course, carry with it an underestimation of the formal features nor detract from the enjoyment of their quality nor imply that real results already achieved in art-historical literature through formal analyses have lost their validity. Rather the contrary.[3] We can foresee that within two or three generations a new overall pattern of stylistic developments will have been evolved. Such a pattern will buttress and clarify the purely formal evolutions already established by pegging them to a basis wider than previously thought possible.

Why is it, we may ask, that a tendency still remains among some art-historians to put a brake upon efforts to broaden art-history by a study of social history? As regards England, since the seventeenth and eighteenth centuries an admirable tradition of art theory, art criticism, and connoisseurship has flourished here. Art-history, on the other hand, as a university

---

[1] *Courbet and the Naturalistic Movement.* Essays read at the Baltimore Museum of Art, edited by G. Boas, Baltimore, 1938.

[2] *Space, Time, and Architecture*, Cambridge Mass., 1941.

[3] The results on re-gothicization during the Quattrocento at which I arrived some twenty-five years ago through formal analyses have now been confirmed through my study of the whole historical material. In a recent article ('Observations on Girolamo da Carpi', see above p. 107) I have also tried to show how the continuation of Quattrocento gothic in Mannerism, which I saw in my earlier writings mainly as a formal process, was ultimately based on the social changes.

discipline, obliged to stand on its own feet, work out its own field of research and its own method is of very recent growth. The new science necessarily originated in previous art criticism; at first, towards the end of the nineteenth century, in its more impressionistic form, art criticism was largely concerned to describe the fleeting reactions of a sensitive beholder before a work of art, while later, in the early twentieth century, an attempt was made to modify this extreme subjectivism by a more controlled, more constructed, but still unhistorical approach. The historical point of view naturally came into the new discipline where it was the most urgently needed, the most obviously lacking, and where a transition from the previous stage of art criticism could be most easily effected: in the construction of the historical development on a formal basis. So the space allotted to history within art-history was relatively small, as it had been in the Wölfflin school. But, while in the Wölfflin school the theory of art for art's sake could only be sensed as a distant though necessary phenomenon, here art-history, because of its later origin in an esoteric art criticism, was still closely and directly bound up with it. It is almost a hundred years since Ruskin, than whom none could have been more averse to the art for art's sake attitude, considered art as expressive of the society which produced it, if mainly of the ethical life of society, and was stimulated in consequence of his study of art to a thorough study of the social structure and social economy.[1] In contrast to Ruskin not only many writers during his later life-time and after him, but even some art-historians of our own day have still been apt to believe, fundamentally, that art is a world by itself which has, and should have, as little contact as possible with the tangible world. Since they cannot be consistently historical, these latter still adhere to the supposition that the art for art's sake point of view is unchangeable. They cannot imagine that art-history is a piece of history[2] and that the art-historian's task is primarily not to approve or to disapprove of a given work of art from his own point of view, but to try to understand and explain it in the light of its own historical premises; and that there is no contradiction between a picture as a work of art and as a

[1] In his Ruskin lecture, *Ruskin's Politics*, London, 1921, Bernard Shaw drew attention to this evolution remarking, incidentally, that this marked his own development too.
[2] I purposely employ the expression 'piece of history' because Saxl (himself a product of the school of Vienna) used it in conversation with me.

document of its time, since the two are complementary. Nor can they appreciate that familiarity with outlook and taste aids us in comprehending, not only the complete style of a picture, but ultimately, even its quality: partly because the quality of a given picture, in its special nuance, can only be seriously judged if compared with other pictures of the same style and even more so because knowledge historically-grounded is the only sure means of neutralizing our subjective judgment on the quality of works of art of the past, even on the significance of individual styles, which otherwise is too exclusively conditioned by our penchant for one tendency or another in contemporary art akin to them.[1]

In recent years, as is well known, historical scholarship in England has tended to emphasize the economic and social aspects. Yet, for instance, though Tawney's book, which we have mentioned, is one of the most widely-read, art-historians of the older persuasion appear to be unacquainted with the fruitful achievements of modern historical research which is to be found, so to speak, on their doorsteps.[2] It is distasteful to them to find, embedded in art-historical literature, facts and terms, commonplaces in every historical book, with which they are unfamiliar and the art-historical implications and consequences of which they fail to grasp.[3] Living in their

[1] Interest in baroque and mannerist art, which originated, as we have mentioned, within the Viennese school of art-history, was a consequence of the growing historical thoroughness of this school's own researches, while at the same time reflecting contemporary art tendencies. The analysis of baroque art began with Riegl in the last years of the nineteenth century, that of mannerism with Dvořák in the years preceding World War I. Though recognition of the qualities of the latter style, of course, coincided with the taste for contemporary expressionist art, as time goes on and with the growth of our knowledge of the spiritual and social background of mannerism, we shall obtain an increasingly objective idea of this style.

[2] Nor, to take an example nearer to art-history, do they appear to be conversant with the writings of the outstanding prehistoric archaeologist, V. Gordon Childe, which extend not only to oriental but even to classical antiquity and which establish the closest possible relations between social structure, religion, mental outlook, and art.

[3] The connection between religious and economic thought, as Tawney has demonstrated, can no longer be disputed today. Yet, it is apparently only the historian who is allowed to be aware of this, not the art-historian, and, if the pure formalists had it their own way, art-history would be destined to carry on in a water-tight compartment cut off from the other historical disciplines. This is even true of terminology. In the 1937 preface to his

ivory tower, they think that to adduce the results of social or ecclesiastical history must degrade an art-history which should, at least theoretically, be reserved to masterpieces and in which the diversity of styles is explained by the diversity of styles. The sensitiveness and esoteric nature of their spiritual ancestors has by now become a search for precious, if possible, unusual words. We can feel no surprise, therefore, under such conditions, if the non-art-historian, in particular the social historian, for example E. Halévy, in the short chapter on art in his *History of the English People*, 1815 (English translation, London, 1924) can make striking, new art-historical observations which, in many ways, are more interesting and revealing than those of some art-historians on the same period.

The whole point of view of art-historians, of whatever country or training, who have not yet even absorbed the achievements of Riegl, Dvořák, and Warburg (let alone tried to go beyond them) is conditioned by their historical place: they cling to older conceptions, thereby lagging behind at least some quarter of a century. And, in the same way are conditioned their step-by-step retreat and the concessions they are willing to make—not too many and not too soon—to the new spirit. Their resistance is all the stronger, their will to give ground, all the less, the greater the consistency and novelty they encounter. They themselves frequently publish weak pictures by fifth-rate masters, provided the period is remote enough: for these are attributions to, say, the Master of the Goodenough Deposition, and thus are justified from the point of view of connoisseurship.[1] Even the abundant literature on popular and semi-popular art is not, I believe, particularly frowned upon

[1] To avoid any misunderstanding: nothing would be more puerile than to deny the obvious importance of attributions. What will soon be gone with the wind is that over-accentuation, which tends to confine art-history to attributions almost for attributions' sake.

book, *Religion and the Rise of Capitalism*, originally published in 1926, Tawney wrote: 'When this book first appeared, it was possible for a friendly reviewer, writing in a serious journal, to deprecate in all gravity the employment of the term "Capitalism" in an historical work, as a political catch-word, betraying a sinister intention on the part of the misguided author. An innocent solecism of the kind would not, it is probable, occur so readily today.' It can, however, occur even in 1948 from the same innocence, when it is a question not of historians but of art-historians, who, as regards certain current terms, are fettered by a primitive word fetishism.

so long as this art is kept well apart from the general stylistic development or, at any rate, can be considered diverting and charming, reminiscent of Henri Rousseau. Discussion of the subject-matter seems permissible as long as it is restricted to an iconography in which the explanation of the choice of subject is kept as aloof as possible from living history. Literature on the working conditions of artists is not, I think, objected to, provided it remains detached and conclusions which could be drawn from it are not incorporated into literature dealing with great artists but are limited to isolated and casual reference. The innumerable allusions in art-historical literature to the social and political background usually pass unchallenged as long as the connection between it and art is left, on the whole, comfortably vague.[1] In the case of some artists of more recent centuries, however, practising secular art, the connection is so obvious that constant reference to it in literature has bred familiarity: in the case, for instance, of Hogarth, David or Géricault. Thus, a step further which reaches the precise association of style and outlook, a step so small that it is scarcely noticeable, passes without comment.[2] But when it is no longer a question of secular art of the eighteenth or nineteenth centuries but, let us say, of religious art of remoter times as was, for instance, the case in my book on Florentine painting of the fourteenth and early fifteenth centuries, then there still appears an objection on the part of some art-historians to the discussion of differences in religious sentiment and consequently in religious art, as associated with various social groups; they would prefer to keep Fra Angelico and Botticelli in the dream-world ambient where the pre-Raphaelites put them. Although lately it has become fashionable to introduce a few historical facts, these may only enter the art-historical picture when confined to hackneyed political history, in a diluted

---

[1] C. Gutkind's *Cosimo de' Medici*, Oxford, 1938, is a typical case where a well-meaning author has felt the need to adduce far more economic and social history than had previously been done, but has not yet arrived at the stage of drawing any conclusions from them or of connecting them with anything. A large part of the book deals in almost too great detail with the economic conditions in Florence and with Cosimo's business interests, while, in the chapter 'Cosimo in Private Life', his philosophy of life (and, of course, also his liking for art and learning) remains entirely detached, so that we acquire no all-round picture of Cosimo's person and outlook.

[2] Articles I wrote on those three artists and to which no exception was taken were in the same vein as my book on Florentine painting, mentioned below.

form, which gives as little indication as possible of the existing structure of society and does not disturb the romantic twilight of the atmosphere. The last redoubt which will be held as long as possible is, of course, the most deep-rooted nineteenth-century belief, inherited from Romanticism, of the incalculable nature of genius in art. It is, however, characteristic of the strength of the new trend that L. Münz, the best connoisseur of Rembrandt in our day, should have brought out, in 1931, a popular, annotated edition of Riegl's famous essay of 1902, on the Dutch Portrait Group; here, without detracting in any way from his grandeur, Rembrandt is treated as a link in a long chain and subjected to an analysis so exact and so instructive as to horrify every supporter of the genius theory.[1]

Methods of art-history, just as pictures, can be dated. This is by no means a depreciation of pictures or methods—just a banal historical statement. But the time will naturally come when the exclusive formalists will generally be recognized as in the rear of art-history, as to-day are the antiquarians and anecdotalists.

[1] I would like to recall here Münz's opinion that a closer understanding of Rembrandt's works is gained by the realization that they are charged with meaning and emotion than by those 'happily now obsolete, aesthetic approaches from which Rembrandt's work was seen either as realism empty of all emotional content or as a magic of light and shade so exalted, so unique and intangible, that all attempts to search for a meaning became irrelevant' ('Rembrandt's "Synagogue" and some Problems of Nomenclature', *Journal of the Warburg and Courtauld Institutes*, III, 1939-40).

# Index

# Index

# Index

# Index

# Index

*Plates*

1  Alessandro Allori: Pearl Fishing

2a  *Above*  Federico Zuccari: Last Judgment. Detail
2b  *Below*  Federico Zuccari: The Hunt

3a *Left* Rosso, engraved by Caraglio: Hercules and Achelaus
3b *Right* Taddeo Zuccaro, engraving: Adoration of the Magi

4    Stradanus: The Mine

5  Zucchi: Circumcision

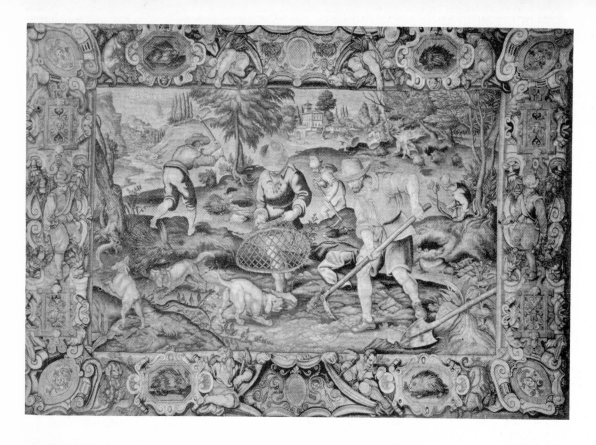

6a  *Above*  Stradanus, cartoon for tapestry: Porcupine Hunting
6b  *Below*  Salviati, Drawing: The Seasons

7a  *Above*  Pontormo: Martyrdom of St. Maurice
7b  *Below*  Candid, drawing: March

8a *Above* Sustris, drawing: Triumph of Galatea
8b *Below* Salviati: Beheading of St. John the Baptist

9a  *Above*  Jacopo Zucchi: Bacchus with attendants. Detail
9b  *Below*  Giovanni di Vecchi. Fresco: Gideon and the Angel

10    Raffaelino da Reggio: Ecco Homo

11 Bartolomeus Spranger: Martyrdom of St. John

12a  *Above*  Wtewael: The Flood
12b  *Below*  Wtewael: Judgment of Paris

13a  *Above*  Aertsen: Crucifixion
13b  *Below*  Floris: Seagods

14   Abraham Bloemart: Niobides

15 Mabuse: Venus

Heu quantum sceleri concedit iniqua voluptas    Obrui immersum cum semel arbitrium.

16a  *Above*  Blocklandt: Joseph before Pharaoh
16b  *Below*  Blocklandt, engraved by Ph. Galle: Temptation of Lot

LOTH EX VNO PERICVLO AB ANGELIS EDVCTVS INSIDIIS FILIARVM SVARVM IN ALTERVM INDVCITVR. GEN. XIX.

17a  *Above*  Floris: Judgment of Solomon
17b  *Below*  Floris engraved by Ph. Galle: Temptation of Lot

18a *Left* Goltzius, engraved by Matham: Christ and the Woman of Samaria
18b *Right* Goltzius, engraving: Honor and Opulentia

19   Martin de Vos: Paul and Barnabas at Lystra

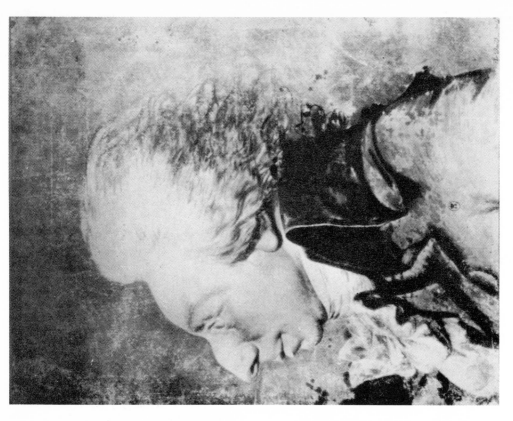

20a  *Left*  David: study for a figure in the Oath of the Horatii
20b  *Above*  David. Drawing: Lepelletier de Saint Fargeau

21 David: The Oath of the Horatii.

22　François André Vincent: Molé and the Partisans of the Fronde

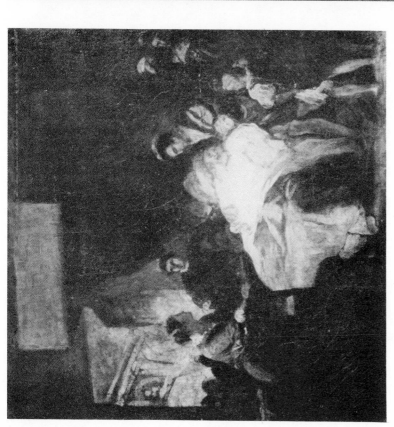

23a   *Left*   Ménageot: Death of Leonardo da Vinci

23b   *Right*   David: Portrait of Barère

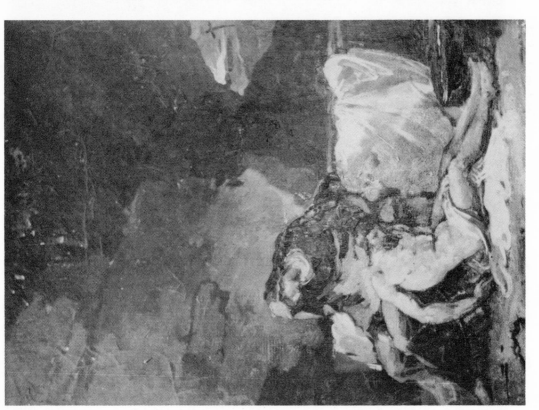

24a  *Left*  Girodet: Pietà.
24b  *Right*  Brenet: The Magnanimity of Bayard

25 Girodet: Atala's Burial

26 Girodet: Ossian receiving Napoleon's Generals

27a *Left* Girodet: Mlle Lange as Danae
27b *Right* François Gérard: Ossian playing the Harp

28a  *Left*   Géricault: White Cat
28b  *Right*  Géricault: Oxen driven to the Slaughterhouse

29a  *Left*  Géricault: Officer of the Imperial Guard
29b  *Right*  Géricault: Cartload of Wounded Soldiers

30a  *Above*  Géricault: Jockeys
30b  *Below*  Géricault: Race of Riderless Horses

31a  *Above*  Géricault: Raft of the Medusa
31b  *Below*  Géricault. Lithograph: Entrance to the Adelphi Wharf

32a  *Above*  Girolamo da Carpi: Adoration of the Magi
32b  *Below*  Géricault. Drawing: Slave Trade

33a  *Above*  Géricault: Le Four à Plâtre
33b  *Below*  Peruzzi: **Adoration** of the Magi.

34a  *Left*  Girolamo da Carpi: Portrait of a Lady
34b  *Right*  Girolamo da Carpi: Madonna. Detail of the Adoration of the Magi

35a *Left* Pontormo: Portrait of a Lady

35b *Right* Parmigianino: Portrait of Antea. Detail.

36a *Above* Alfonso Lombardi. Relief: Adoration of the Magi
36b *Below* Battista Dosso and Girolamo da Carpi: one of the Horae with the Steeds

37a  *Above*  Peruzzi. Fresco: Ganymede
37b  *Below*  Girolamo da Carpi: Ganymede

38a  *Left*  Girolamo da Carpi: Chance and Penitence

38b  *Right*  Girolamo da Carpi: Pentecost

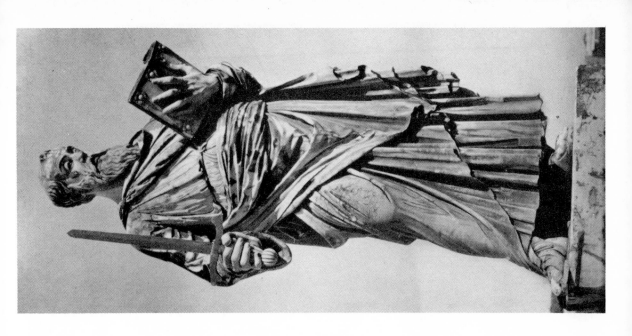

39a *Left* Girolamo da Carpi: Studies in pen and ink
39b *Right* Paolo Romano. Sculpture: St. Paul

40    Garofalo: Triumph of Bacchus

*Opposite*

41*a*    *Above*    Girolamo da Carpi. Pen, ink and wash: copy after Garofalo's
            Triumph of Bacchus
41*b*    *Below*    Girolamo da Carpi. Fresco: Triumph of Bacchus

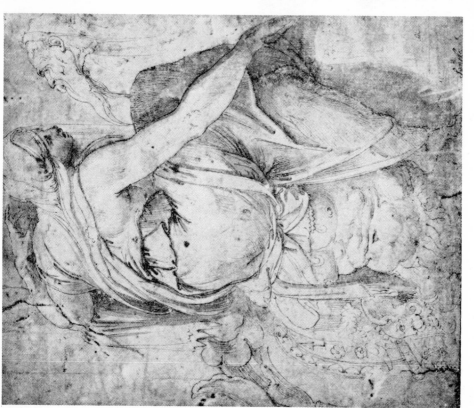

42a  *Left*   Franco. Pen, ink and wash: Seated Woman
42b  *Right*  Franco. Fresco: Detail. Capture of the Baptist

43   Girolamo da Carpi. Pen and light brown ink: Holy Family

44a *Above* Girolamo da Carpi. Pen and brown ink: copy after the Antique

44b *Below* Nicolo dell'Abbate. Wash: Concert

45   Nicolo dell'Abbate: Portrait of a Lady

46  Joseph Highmore: Mr Oldham and his Guests

47a  *Left*  Joseph Highmore: Self-portrait
47b  *Right*  Joseph Highmore: Samuel Richardson

48a  *Left*  Michele di Ridolfo, here attributed to: Portrait of a Young Man

48b  *Right*  Michele di Ridolfo: Lucretia